Gerry Stahl's assembled texts volume #1

Marx & Heidegger

Gerry Stahl

Gerry Stahl's Assembled Texts

- *Marx and Heidegger*

- *Tacit and Explicit Understanding in Computer Support*

- *Group Cognition: Computer Support for Building Collaborative Knowledge*

- *Studying Virtual Math Teams*

- *Translating Euclid: Designing a Human-Centered Mathematics.*

- *Constructing Dynamic Triangles Together: The Development of Mathematical Group Cognition*

- *Essays in Social Philosophy*

- *Essays in Personalizable Software*

- *Essays in Computer-Supported Collaborative Learning*

- *Essays in Group-Cognitive Science*

- *Essays in Philosophy of Group Cognition*

- *Essays in Online Mathematics Interaction*

- *Essays in Collaborative Dynamic Geometry*

- *Adventures in Dynamic Geometry*

- *Global Introduction to CSCL*

- *Editorial Introductions to ijCSCL*

- *Proposals for Research*

- *Overview and Autobiographical Essays*

- *Theoretical Investigations*

- *Works of 3-D Form*

Gerry Stahl's assembled texts volume #1

Marx & Heidegger

Gerry Stahl

Gerry Stahl

Gerry@GerryStahl.net

www.GerryStahl.net

Published by Gerry Stahl at Lulu.com

Printed in the USA

ISBN 978-1-329-85660-8 (paperback)

ISBN 978-1-329-85650-9 (ebook)

Introduction

T his volume contains the doctoral dissertation of Gerry Stahl in Philosophy at Northwestern University. It was entitled: "Marxian Hermeneutics and Heideggerian Social Theory: Interpreting and Transforming Our World." The dissertation was defended on May 8, 1975.

The original typewritten version was scanned and an electronic version was created on the 25th anniversary of the document. Changes were limited to minor stylistic improvements and the graphics. Digital copies are available in html and pdf format at:

http://GerryStahl.net/publications/dissertations/philosophy.

Additional graphics have been added for the 35th anniversary edition.

The dissertation considers the two most important philosophers of the modern age. When I was there, Northwestern had the only American department of philosophy that encouraged the study of European philosophy. I also conducted my research during three years in Germany: at Heidelberg, where Heidegger's work was continued, and at Frankfurt, where critical theory extended Marx' thinking.

In recent years, I have applied conceptual and methodological perspectives from Marx and Heidegger to the theory of CSCL. In particular, Marx countered the ideology of individualism by analyzing social structures and interpersonal interactions at different units of analysis than the individual person. Heidegger also questioned the traditional ontology of natural objects with innate attributes by proposing dynamic interactive processes of beings in their ecological context.

Today, the philosophies of Marx and Heidegger are still extremely relevant—provided one adapts them to the current socio-historical context and adjusts each to the implicit criticisms of the other.

Marxian Hermeneutics and Heideggerian Social Theory:

Interpreting and Transforming Our World

by

Gerry Stahl

a dissertation

submitted to the Graduate School

in partial fulfillment of the requirements

for the degree

Doctor of Philosophy

Northwestern University

Evanston, Illinois

June 1975

The original typewritten version is available at Northwestern University:

Diss

378

N.U.

1975

5781m

It is indexed in Dissertation Abstracts and available from University Microfilms

Abstract

**Marxian Hermeneutics and Heideggerian Social Theory:
Interpreting and Transforming Our World**

Gerry Stahl,

Northwestern University

June 1975

Today neither philosophy of interpretation (hermeneutics) nor philosophy of society can legitimately proceed without the other. Interpretation of the world precedes the possibility of transforming it, according to Martin Heidegger, because the presence of beings is always already meaningfully structured. For Karl Marx, however, interpretations of the world are constituted by human *praxis*, the reproduction and transformation of social reality. The confrontation of Marx's thought with Heidegger's provides an appropriate historical medium for the indispensable task of bringing the problematics of critical social theory and philosophical hermeneutics to bear upon each other.

The alternative notions, that hermeneutics either founds or is founded upon social analysis, are reconciled by interpreting Marx's social methodology as being in accord with hermeneutic principles and by transforming Heidegger's ontology to take account of social mediations. Thereby, Heidegger's critique of metaphysics clarifies Marx's methodological sophistication, rescuing Marxism from a history of mechanistic corruptions, while Marx's insights into the power of social relations provide a corrective to the politically reactionary self-understanding, abstract form, scholastic structure and non-social content of Heidegger's jargon.

Thinking about Marx and Heidegger together is most fruitfully accomplished by a sympathetic study of their mature approaches and systems, focusing on the relation between beings and Being, the concrete and the abstract, the individual entity and its socio-historical context. Hermeneutic, political and internal justifications for the selection of specific primary texts, for not making explicit use of secondary works, and for interpreting the two philosophers through each others' eyes are indicated in the introductory Part I. Above all, it is argued, a contemporary perspective on Marx is inevitably affected by Heidegger's influence as well as by intervening political developments; and similarly for reading Heidegger.

Marx's theory of commodity fetishism plays a role analogous to Heidegger's theory of the oblivion of Being. In both systems, the distorted appearance of things is related to the prevailing form of the Being of beings: their commodity form for Marx or their technological character for Heidegger. The commodity form of products and of human productive labor prevails in the bourgeois or capitalist era. Marx, whose methodology is specific to an analysis of this period, traces the historical and structural development of these commodity relations in primarily socio-economic terms. The way in which changes in the over-all social character are thereby related to concrete interactions provides the guiding theme of the Marx interpretation, which forms Part II.

Where Marx relates the technological character of the commodity to its actual, concrete, everyday exchange in the marketplace as historically developed, Heidegger insists that the process by which, e.g., the technological character of beings has been given, the *"Ereignis,"* is ungrounded and incomprehensible. But such an insistence ignores the proper position of the *Ereignis* within Heidegger's system: as the process of self-mediation and of totalization of all that which is present, The analogy between the role of the social character in Marx's system and that of the *Ereignis* in Heidegger's is drawn in the opening and closing remarks of Part III, the Heidegger interpretation. There it is argued that Heidegger's alternative conceptualization weakens Marx's sense of the historical limits of theory as well as foregoing all ability to comprehend transformations of Being or society concretely.

Considering Heidegger and Marx together suggests that Heidegger's central fault is in failing to relate changes in Being – the historically prevalent form of presence of beings – to developments within the concrete social realm of entities. Changes of ontological interpretation can, as Marx demonstrates, be comprehended in terms of transformations within society, whereby, of course, the social theory must itself be hermeneutically appropriate.

Marxian Hermeneutics and Heideggerian Social Theory:

Interpreting and Transforming Our World

Man müsse durch die Eiswüste der Abstraktion hindurch, um zu konkretem Philosophieren bündig zu gelangen.

–   Adorno quoting Benjamin

Preface

Today neither philosophy of interpretation (hermeneutics) nor philosophy of society can legitimately proceed without the other. Interpretation of the world precedes the possibility of transforming it, according to Martin Heidegger, because the presence of beings is always already meaningfully structured. For Karl Marx, however, interpretations of the world are constituted by human *praxis*, the reproduction and transformation of social reality. The confrontation of Marx's thought with Heidegger's provides an appropriate historical medium for the indispensable task of bringing the problematics of critical social theory and philosophical hermeneutics to bear upon each other.

The alternative notions, that hermeneutics either founds or is founded upon social analysis, are reconciled by interpreting Marx's social methodology as being in accord with hermeneutic principles and by transforming Heidegger's ontology to take account of social mediations. Thereby, Heidegger's critique of metaphysics clarifies Marx's methodological sophistication, rescuing Marxism from a history of mechanistic corruptions, while Marx's insights into the power of social relations provide a corrective to the politically reactionary self-understanding, abstract form, scholastic structure and non-social content of Heidegger's jargon. Such a consideration of Marx and Heidegger together strengthens the position of each. Because they stand firmly within a shared post-Hegelian German tradition, the merging of their ideas proceeds by merely drawing out what is already implicitly present.

Thinking about Marx and Heidegger together is most fruitfully accomplished by a sympathetic study of their mature approaches and systems, focusing on the relation between beings and Being, the concrete and the abstract, the individual entity and its socio-historical context. This strategy determines the selection of texts to be analyzed. Rather than centering on accidentally parallel discussions of explicitly political issues, writings are chosen with the goal of developing the most important systematic and methodological themes of Marx's and Heidegger's thought. Their mature presentations – Volume I of *Das Kapital* (1867) and the lecture on *Time and Being* (1962) – are taken as standards, with other works drawn upon to trace the developments leading up to them. Hermeneutic, political and internal justifications for the selection of specific primary texts, for not making explicit use of secondary works, and for interpreting the two philosophers through each others' eyes are indicated in the introductory Part I. Above all, it is argued, a contemporary perspective on Marx is inevitably affected by Heidegger's influence as well as by intervening political developments; and similarly for reading Heidegger.

While less central points of direct contact between the writings of Marx and those of Heidegger have been ignored, several correspondences have been thematized. A primary motivating presupposition of both Marx's and Heidegger's project is the belief that true reality lies hidden from our direct perceptions. Marx's theory of commodity fetishism plays a role analogous to Heidegger's theory of the oblivion of Being. In both systems, the distorted appearance of things is related to the prevailing form of the Being of beings: their commodity form for Marx or their technological character for Heidegger. Heidegger's "technological stock" has essentially the same characteristics as Marx's "commodity." Both forms are, furthermore, historically specific. Technological stock is the characteristic form of the Being of beings in the modern epoch, which is, according to Heidegger, historically given by Being-as-such or the *Ereignis*. Correspondingly, for Marx, the commodity form of products and of human productive labor prevails in the bourgeois or capitalist era. Marx, whose methodology is specific to an analysis of this period, traces the historical and structural development of these commodity relations in primarily socio-economic terms. The way in which changes in the over-all social character are thereby related to concrete interactions provides the guiding theme of the Marx interpretation, which forms Part II.

Where Marx relates the technological character of the commodity to its actual, concrete, everyday exchange in the marketplace as historically developed, Heidegger insists that the process by which, e.g., the technological character of beings has been given, the "*Ereignis*," is ungrounded and incomprehensible. But such an insistence ignores the proper position of the *Ereignis* within Heidegger's system: as the process of self-mediation and of totalization of all that which is present. To divorce mediation from its content is hypostatization; to project social totalization beyond its socio-historical limits is to fall behind Marx's level of methodological self-reflection. The analogy between the role of the social character in Marx's system and that of the *Ereignis* in Heidegger's is drawn in the opening and closing remarks of Part III, the Heidegger interpretation. There it is argued that Heidegger's alternative conceptualization weakens Marx's sense of the historical limits of theory as well as foregoing all ability to comprehend transformations of Being or society concretely.

Considering Heidegger and Marx together suggests that Heidegger's central fault is in failing to relate changes in Being – the historically prevalent form of presence of beings – to developments within the concrete social realm of entities. Changes of ontological interpretation can, as Marx demonstrates, be comprehended in terms of transformations within society, whereby, of course, the social theory must itself be hermeneutically appropriate.

***

The methodological reflections on thinking about Marx and Heidegger together, the interpretation of Marx, and the analysis of Heidegger are each carried out in three chapters, as summarized below:

The dialectic of essence and appearance at work in the systems of both Marx and Heidegger represents a shared response to present social appearances as obscuring the potential for a better world, one which would incorporate new forms of ontological relations (Part I). But the two mainstreams of contemporary continental thought which flow from these systems, and which appeal especially to those interested in transforming the world, problematize each other. Issues both internal and external to Marx's theory and Heidegger's thought call for a reckoning by each with the other (Chapter I). Heidegger, for instance, accuses Marxism of adopting "metaphysical" conceptualizations (Chapter II), while Marxists respond that Heidegger has ignored the impact of social conditions upon his thought (Chapter III).

Marx's works are construed as interpretations of the social relations underlying appearances which have been distorted by capitalist relations (Part II). His early writings, *Alienated Labor* and *Theses on Feuerbach*, anticipations of his mature critique of political economy, occasionally substitute the critical appropriation of prevalent metaphysical hypotheses for the stringent methodology subsequently used (Chapter IV). Marx's *Grundrisse* then develops the appropriate historical analyses, economic categories and hermeneutic methodology though theoretical research (Chapter V). Finally, *Capital* systematically presents the analysis of capitalist society, starting dialectically from the abstractions arrived at in the capitalist economy (Chapter VI). The hermeneutic accord between Marx's interpretations of the world and the historic processes which reproduce and transform the world, the manifold unity of Marx's social theory and capitalist social practice, saves Marx's system from the charge of being metaphysical by deriving its method from its object.

Heidegger's post-war thought offers an alternative to Marxism by focusing on the general, non-economic relationship between entities and their form of presence in a given historical epoch (Part III). *The Origin of the Work of Art* presents Heidegger's "reversal" toward Being-as-such, formulating his central question of Being in terms of the origin of the historically specific form of presence of a work which establishes its own presence (Chapter VII). The tendency here to give an absolute priority to Being develops in the essay *The*

Thing, which introduces his mature theoretical framework. (Chapter VIII). Heidegger's final statement, the lecture on *Time and Being*, takes a meta-ontological overview of the history of the forms of presence which, however, leaves the concrete details of historical ontological transformations shrouded in mystery (Chapter IX). Thereby, the ontological self-interpretation of the world is illegitimately divorced from its ontic self-transformation, leaving Heidegger's social commentary content-less and messianic next to Marx's.

***

Note: Chapter III is copywritten by the journal in which it appeared as "The Jargon of Authenticity: An Introduction to a Marxist Critique of Heidegger" by Gerry Stahl *(Boundary II*, Department of English, SUNY-Binghamton, NY 13901, Winter 1975, pp. 439-497).

Quotations: All quotations are given in English. Translations from the German are based upon the best available English versions, but are revised without notice for increased literalness and consistency. References to texts of Marx and Heidegger are given to both the translation and the original, with English page numbers preceded by *p* and German by *S*.

***

The present work represents the culmination or thirty years of progress toward the author's intellectual maturity. As such, it is a token of gratitude to all those who have contributed, however unknowingly, to that process. It is, accordingly, *dedicated to those magical moments when truth makes its appearance unannounced, but deservedly, within a social gathering.*

Contents

PART I: INTERPRETING MARX AND HEIDEGGER IN OUR WORLD

The author thinking about Marx and Heidegger during a visit to East Berlin in 1972.

Chapter I. The Alternative of Marx and Heidegger

The reasons for my decision to write on Karl Marx and Martin Heidegger together are numerous. Throughout my study of philosophy, the two major tendencies in continental thought, Marxism and existentialism, have been rivals for my interest. Marx and Heidegger are clearly the founders of the two schools and to my mind they remain the most profound representatives. It was thus natural that I should take the opportunity of researching a dissertation to come to grips with the philosophical alternative they present.

My personal inclination is not, however, merely subjective; it is an expression of the objective conditions in society and in the philosophical tradition. There are, that is, good reasons for someone critical of today's society to be repelled by the inherent conservatism of Anglo-American philosophy and to be attracted to Marx and Heidegger. Both Marx and Heidegger, for all their criticisms of Hegel, retain the central insight of dialectics: that the facts are not simply given, but are mediated in ways that can only be comprehended with the help of theory. A philosophy that does not take this seriously is ill equipped to deal with deceptive reality.

To turn to Marx or Heidegger as to a dogma is, however, to destroy them. Not only does the originality of their thought demand an intellectual struggle that critically overcomes the habits of common sense, but the weaknesses which have become apparent in their systems necessitate creative development of these systems. Internal requirements of the two philosophies, as well as their deficiencies, call for a confrontation between them, which could serve to clarify and strengthen each of the alternatives, if not to synthesize them. The present introductory chapter and the subsequent review of previous debates between the two positions outline these needs, anticipating the material which follows in the actual interpretations.

A basis for comparison of the two approaches exists in terms of the common search for essences hidden in appearances. The differences between the essential concepts they form and emphasize suggest, then, that Heidegger can be understood as a rethinking of Marx, who too narrowly based his analysis on the economic realm. On the other hand, the lack of historical content in Heidegger's concepts needs to be remedied through a study of Marx's method of historically-specific concept formation. Although a review of previous attempts at interpreting Marx and Heidegger from each other's perspective reveals that there has been little success to date in this enterprise, previous misunderstandings can generally be attributed to national and international politics, and it can be hoped that a more fruitful dialogue is now possible.

Chapter I concludes with a summary of the themes and considerations which are raised in Part I and which determine the outlines of the subsequent interpretations of Marx and Heidegger. Chapters II and III expand upon the comparison of Marx and Heidegger by reviewing Heidegger's critique of Marx and Adorno's Marxist critique of Heidegger. These chapters thereby uncover internal arguments why Heidegger should have paid more attention to Marx and why Marxists must come to terms with Heidegger's thought.

Interpretation for Transformation

There is today a need for interpretation of the world. Marx and Heidegger share with Freud the belief that it is possible with the help of a theory to understand someone's ideas, behavior and self better than he understands them himself. The motivations consciously debated by the agent may well be screens against true perception or at best interpretations of his situation, which are not necessarily privileged over the analysis of his situation by other people. The idealistic presupposition of the transparency of the *cogito* to the *ego* has been rejected by these post-Hegelian outlooks. The subject, who has been raised in a family, mediated by social conditions, and "thrown" into the world, must interpret his own consciousness, activity, and Being just as an observer must, namely from a perspective which may well be more limited by ignorance of various factors and by being more caught up in self-concealing conditions than an observer with a developed theory – even though the subject has been exposed to more of the empirical facts. This is not a merely scholastic question of epistemology. The self-perception of the subject situated naturally (i.e., without the objectifying alienation of theoretical analysis) in his family, society and world is in fact subject to systematic distortions of which he remains unaware. The normal psychic dynamic of family life is predicated upon its sublimation into the unconscious; the invisible hand of bourgeois exchange society could not be effective without commodity fetishism; and the reliability of the world presupposes that we are "fallen" in it and do not recognize its "worldhood" or "worlding," its Being.

Both Marx and Heidegger situate Hegel's dialectic of essence and appearance in the contemporary world. Marx argues that capitalist society is pervaded by a "fetishism of commodities," that is, that the essential social relationships which structure society and the lives of its members appear, if at all, in the illusory form of characteristics of physical objects, of the commodities produced. Any evaluation of capitalist society in terms of its appearances alone, without the assistance of a theory which interprets and demystifies the appearances will necessarily be apologetic – at most liberally reformist – covertly and dogmatically endorsing the mystifying ideology of capitalism. A theoretical interpretation of the essences as illusion, on the other hand, allows for a critical grasp of their contradictory nature and reveals potentials for qualitative transformation.

Similarly, Heidegger argues that Western thought is guilty of a progressive "forgetfulness of Being" such that the ontological categories through which we understand reality distort our relationships to ourselves and other beings. What is needed is a meta-ontology, a theory which deals with the deceptive character of contemporary appearances. Thus, common to Marx and

Heidegger, but not to the competing philosophic approaches of the twentieth century, is the belief that appearances by themselves are illusory, the insight that this illusory character is historically situated, and the conviction that philosophy's task is to break through such illusion. This shared conviction provides a basis for the following interpretations of Marx and Heidegger and for their comparison. The central methodological problem for both thinkers is accordingly the question of how to derive the appropriate theoretical essence from the given appearances, from the ideologies and the phenomena. The different approaches to a shared project determine contrasts between Marx and Heidegger, which are clear in their respective conceptual frameworks, or rather, in the way in which they try to avoid imposing conceptualizations external to their subject matter.

Marx and Heidegger each formulate an essential concept. Marx raises the question, What is truth? by arguing that capitalist appearances are illusory, fetishized, false. This alone might qualify him for consideration as a philosopher in the broad sense of a thinker who stops at no academic borders. Frequently, however, he is relegated to the ranks of out-moded economists. Worse yet, perhaps, his thought is accepted as interdisciplinary, and segmented according to the academic division of labor against which it stands as a forceful counterexample. A preferable way of understanding the complexity of Marx's thought is suggested by Jürgen Habermas' analysis of *emancipatory* science as a dialectical unity of *interpretive* and *explanatory* interests.[1]

Speculative philosophy (of the Hegelian tradition) is concerned to *interpret* reality, to provide categories for subsuming reality such that the system of categories provides a sense or meaning in terms of which reality can be understood, comprehended, interpreted. Such philosophy is retrospective, not predictive; it does not make calculations, but interprets the significance of their results. Non-dialectical philosophy and science are *explanatory* in the sense that they construct their concepts operationally, formulate laws to predict in quantitative terms, clarify logical difficulties and anomalies. They are thus useful for manipulating events within the given norms, but inadequate by themselves for criticizing these norms. Because Marx wants both to comprehend reality critically and to explain its functioning and its development with an eye to transforming it, his theory must be both interpretive and explanatory.

---

[1] Jürgen Habermas, "Knowledge and human interests: A general perspective," *Knowledge and Human Interests*, translated by Jeremy J. Shapiro (Boston: Beacon, 1971). Cf. Jürgen Habermas, "Erkenntnis und Interesse," *Technik und Wissenshaft als Ideologie* (Frankfurt: Suhrkamp, 1968).

To understand Marx is to comprehend the unity of these two aspects of his work. Nevertheless, one can roughly say that Marx's theory of value (in *Capital*, Volume I) is primarily interpretive (of the essence), while his price theory (*Capital*, Volume III) is primarily explanatory (of the appearance). We shall be concerned with Marx's interpretive framework, rather than with his explanatory science. The criticisms which the technical details of the latter have received by even Marx's most sympathetic readers is not the least reason for reconsidering Marx's interpretive theory in relation to present society and in comparison to competing philosophies. The mediation of Marx's value theory with his price theory – which gives the unity of interests to his critical theory of society – takes place in terms of the consideration of more and more economic influences. The starting point for the entire system is the *commodity*, cornerstone of capitalist production. The theory of capitalist society, including the analysis of fetishism, which is the basis of the critical thrust of Marx's system, can be presented by unpacking this abstract concept. For Marx's concept of the commodity summarizes the results of years of social research and theoretical critique which he dedicated to developing his early, anticipatory social criticisms.

Despite the fact that many social critics today feel that Marx's systematic focus was too narrowly economic, surprisingly few alternatives to Marx's approach have been developed. Either Marx's theory is patched up or research into delimited realms of appearance is carried out with little theoretical guidance. Martin Heidegger's thought suggests itself as a broader alternative to Marxism. His philosophical theory is not only *prima facie* comparable to Marx's, but in many respects methodologically quite similar. Furthermore, there are historical reasons for viewing this alternative as a rethinking of Marxism. Heidegger's mature thought can well be understood as the attempt to interpret reality, including its illusory character, more radically than Marx by reflecting upon the ontological categories at work in capitalist production and more generally in our modern age. In his theory, the concept of *technological Being* plays roughly the same role as that of the commodity in Marx's. Two crucial questions in evaluating Heidegger's alternative to Marx are: Has Heidegger really thought about Marx adequately, that is, has he understood the significance of Marx's accomplishments? Secondly, has Heidegger really been more radical than Marx or has he in fact fallen behind Marx's standpoint philosophically as well as in terms of content? These questions are to be understood quite apart from the undisputed fact that Heidegger's theory is not as fully developed in concrete details as Marx's, that Heidegger has, by his own admission, just managed to clear the ground somewhat.

The concepts of a critical theory of society are perforce radically historical. They display a temporal structure all their own. If the given appearances are illusory, then the concepts that name them effectively must be able to move dialectically between essence and appearance. In temporal terms, the concept must show that appearances lack necessity, that the past was essentially different. As critical, the concept also proclaims the possibility of a better future; it anticipates a qualitative transformation.

Marx's key concept, that of the commodity, is not limited to the era which it characterizes. Nor is it simply universal. Rather, it can retrospectively shed light on its less developed forms under feudalism and also suggest the form it might take in a subsequent harmonious industrial society. Briefly, that is, the relation between the two primary moments of the commodity, use value and exchange value, mirrors the historically changing tensions within society as a whole, their relation of opposition within the capitalist form of production had not yet developed before capitalism and would have to be overcome in the future in order to transcend fetishism, alienation, exploitation, and impoverishment. Within Heidegger's system, much the same can be said about the technological character of Being. In his terms, it is the "Janus head" facing both danger and salvation, one foot in the present epoch and another in a possible subsequent one. Retrospectively, it also makes sense of the development that led up to it.

For a theory to move between essence and appearance, to interpret the development up to the present and to uncover potentials for the future in the present, its key concepts must be neither operationally defined in terms of the given nor ahistorically general. This accounts for the extensive concern with history evident in the work of both Marx and Heidegger. That Heidegger's concepts often seem to lack the historical content characteristic of Marx's suggests that a comparison of the two philosophies may help remove Heidegger's greatest weakness.

Interpreting Marx and Heidegger Together

The problematics of Marx and Heidegger are comparable in fundamental ways. Central to both are the twin paradoxes: guided by *theory*, the analyses must nevertheless be *immanent* to their object; consciously *situated* in the world they interpret, their task is to transform it through *critique*. The unity of critical theory and situated immanence common to Marx and Heidegger defines the tangential point of ideology critique and destructive hermeneutics, social theory and social *praxis*, interpretation and transformation of the world.

Marx and Heidegger follow a *theoretical* approach by focusing on an essential category. This essence, which is elaborated into a conceptual framework, is not simply a concept from which one could logically or dialectically deduce a system, nor does it represent some one being which grounds all other beings as God did in medieval theologies. The theoretical approach is a consequence of the claim that the true structure of reality has been obscured. That this claim does not itself lead to mysticism is due to its being *situated* in the character of capitalist commodity relations or technological Being. Marx and Heidegger see the root of obfuscation in historical developments and strive for the removal of the prevalent deception rather than for submission to it or exploitation of it for purposes of domination. Recognizing the historical objectivity of false appearances, they view their own theoretical insights as moments in the historical transformation required to remove the deceptive character of reality's contemporary self-interpretation. This sense of historical objectivity distinguishes Marx and Heidegger from vulgar utopianism which dreams up ideal societies without concern for making the transition from today's problems. Yet the two thinkers are *critical* in the sense of orienting their thought toward a qualitatively different future. As situated, their theoretical and critical approach is *immanent*. Their orientation toward the future is based on their position in the present, which they understand as having developed out of the past. The character of the systems of Marx and Heidegger, including their methodologies, is explicitly immanent to their historical situation. The theories are articulations of their own circumstances, rather than attempts to impose an abstract, unrelated, ahistorical conceptual framework upon the given. The given is criticized in accordance with its own claims.

However, despite these at least formal similarities, Marx and Heidegger have generally been considered to be at logger-heads. Followers of Marx and Heidegger have maintained primarily polemical relations with each others and previous attempts to think about Marx and Heidegger together have been problematic at best. Since the publication of Heidegger's *Being and Time*, Marxists have dealt with Heidegger in basically two ways. Some, like the early Marcuse or the late Sartre, sought in Heidegger's approach a new ontological foundation or philosophy of man to supplement the analyses of a Marx who supposedly had little time for epistemological questions. Others, like Lukacs, lumped Heidegger's writings in with bistro existentialism and rejected the whole as bourgeois ideology. Generally, the polemicists have been quick to attack surface features without understanding their role in a system that admittedly was until recently only available in the form of obscure hints. Heidegger's apologists, at the other extreme, try to remove all danger of criticism by insisting that he must be interpreted – an endless and thankless task – before he can be judged.

Commentators who have focused on Heidegger's later works have frequently expressed the feeling that Heidegger's thought, for all its depth and breadth, is in the end somehow empty. However, when not hurtled as a weapon of polemic, this objection generally appears camouflaged in the guise of a personal aside tacked onto the end of an objectively argued, uncritical exposition with no attempt to explain the emptiness in terms of what was analyzed. How does this emptiness arise from Heidegger's approach? Where can the problem be pinpointed in his system? What are the ideological implications? What remains of value? The answers to these questions must be sought in the innermost recesses of Heidegger's system. Such a search differs as much from the last minute posing of general "critical" doubts at the end of an uncritical analysis as from an emotional response to surface features. The massive secondary literature on Heidegger seems to lack such a critical search of his system, judging its claim to relevance on the basis of its underlying outlines.

The two knowledgeable attempts to deal with Heidegger as a social thinker fail not only in their over-zealous defense and acceptance of Heidegger's pronouncements, but, more seriously, in seeking something that is not there, Heidegger's "political philosophy" in the Aristotelian sense. Otto Pöggeler's *Philosophie und Politik bei Heidegger* [2] – apparently an attempt to deal with the basic critical problem avoided in his larger commentary on *Der Denkweg Martin Heideggers* – collects many of the central issues and provides a counter-balance to the polemics, without, unfortunately, finding time to go beyond making plausible his defenses of Heidegger. He emphasizes the problem of developing a "political philosophy" on a Heideggerian foundation, without trying to understand how Heidegger's approach already represents an alternative to Marxism.

Alexander Schwan, in his *Politische Philosophie im Denken Heideggers*,[3] tries to adapt Heidegger's analysis of the ontological structure of the work of art simplistically to an analysis of the Hitler state, rather than seeing the art analysis as itself already a social analysis. The arbitrary nature of Schwan's approach becomes striking when he repeats the adaptation with a very different later Heideggerian model with almost identical results. An alternative approach to an analysis of the relation of politics to Heidegger's thought suggests itself in the material on the 1930's which Schwan has himself assembled: to trace the effects of the political climate upon Heidegger's writings or to oppose an analysis of the political phenomena to

---

[2] Otto Pöggeler, *Philosophie und Politik bei Heidegger* (Freiburg: Alber, 1972).

[3] Alexander Schwan, *Politische Philosophie im Denken Heideggers* (Köln: Westdeutscher, 1965).

Heidegger's conception of history – lines of politically critical analysis which are, unfortunately, absent from the political philosophizings of Pöggeler and Schwan.

While few have succeeded in relating Heidegger to Marx, there is an increasing tendency to focus on his similarities to Hegel. Heidegger himself has become more concerned with Hegel in his later writings and seminars, although even *Being and Time* discussed Hegel's conception of time at some length. Heidegger, however, intends to go beyond the tradition that stretched from Plato to Hegel. Hence Kierkegaard, Nietzsche and Marx, the great Hegel critics, are important to him. The concern with Kierkegaard, who allegedly remained on an ontic level, diminished after *Being and Time*, while Nietzsche assumed a central importance in Heidegger's work. After his fascination with Nietzsche waned, Heidegger seems increasingly to have recognized the importance of Marx's post-Hegelianism, without, however, dealing in any depth with Marx. Rather, Heidegger' s references to Marx suggest that a discussion between them is one of the great unfinished tasks of Heidegger's project. An analysis of these references indicates, further, that a necessary first step is to correct the misunderstandings that they express.

The work of Theodor W. Adorno contains a serious and extended critique of Heidegger's system. However, Adorno avoids a treatment of Heidegger's philosophy in isolation. For him, as a Marxist, it is important to deal with Heidegger the way Marx dealt with Hegel: as an expression of the latest stage in the history of philosophy and society. Heidegger's popularity is to be understood in social terms and its ideological consequences are to be combated. Consequently, Adorno's analysis is difficult to judge on a purely philosophical level. Further, while it makes several fundamental points, its form of presentation suffers from abstractness: distance from the material. Not only is the Marxist alternative to Heidegger kept on an implicit level; the interpretation of Heidegger's system remains itself between the lines. Only when supplemented by a thorough interpretation of Marx and Heidegger can Adorno's claims be evaluated, demonstrated, criticized or expanded upon. Particularly bothersome in Adorno's discussions is the way in which he ranges across Heidegger's writings without admitting that they have developed under the recognition of many of the same immanent criticisms which Adorno articulates. Thus, it is useful to focus on one stage of Heidegger's path of thought – his final system – in order to determine just which of Adorno's accusations hold in the end.

The Hermeneutic Context

Heidegger's attitude toward Marx suggests that he has rather uncritically accepted certain prevailing reductionist interpretations of Marx's writings and has thereby reinforced their popularity (cf. Chapter II below). Soviet orthodoxy has not only reified Marx's critical, dialectical thought into a metaphysics, but has used it as a justification for totalitarianism. In rejecting Soviet Diamat, Heidegger (at least until after the war) thinks he is dispensing with Marx, thereby accepting orthodoxy's false claim to authenticity while ignoring what truth may yet be contained in its system. Here, as elsewhere, Heidegger's jargon of origin-al thinking comes into conflict with his insight into the need for "destructive' thought, which starts out from available philosophies to uncover what truth is buried within them. Thus, Heidegger makes a blanket rejection of the economism of Marx as seen through the eyes of the old left (Marxist-Leninists and Social Democrats alike) without considering Marx's arguments for the primacy of commodity production in interpreting our world and, thereby, without being able to up-date the theory to more contemporary needs. Because he does not see the mediation of Marx's economic studies with his philosophy (i.e., his explanatory with his interpretive theory), Heidegger is forced to an extreme humanist interpretation when he wants to salvage something of Marx's thought. By focusing his attention exclusively on Marx's early work as divorced from *Capital*, Heidegger inevitably arrives at the kind of humanist or even existentialist picture of Marx which is so popular in liberal theological circles and which allows him to reject Marx as metaphysically humanist.

In opposition to Heidegger's emphases, the following interpretation of Marx (in Part II) attempts to make sense of his thought as a whole precisely by steering clear of possible metaphysical, economist and humanist distortions in order to arrive at a position that can speak to Heidegger with strength, relevance and independence. Within the context of a presentation of the core of Marx's system, focus will be on Marx's principle of the primacy of commodity production, the unity of his social theory and capitalist social practice, and his analysis of fetishism. It is hoped that the discussion of these focal points will contribute to thought on these important matters. Although the view of Marx presented is conceived as a synthesis of contemporary independent Marxist exegesis, the attempt to structure an interpretation of Marx in terms of the confrontation with Heidegger is, it seems to me, unique and fruitful.

The interpretation of Heidegger (in Part III) follows similar guidelines. The manifold debates over existentialism and Marxism are indicative of the tact that Marxists almost always consider Heidegger an existentialist That is,

Heidegger's doctrine of man in his *Daseinsanalytik* is interpreted moralistically, or at least is taken as an end in itself, as a subjectivistic, individualistic philosophy, rather than as a first step in the anti-subjectivistic questioning of Being. This understanding of Heidegger has not led to significant results because, I suspect, the "existentialism" in *Being and Time* is a popular, superficial level of meaning which merely obscures Heidegger's own thought as developed in his later writings. Adorno's critique is, I think, convincing in arguing that the appealing elements of Heidegger's *magnum opus* are jargonistic and wholly inconsonant with Marxist concerns. The following interpretation thus turns to the late Heidegger, where the accent is no longer on the individual, avoiding, however, the theological interpretation to which Heidegger's ambiguity carefully leaves itself open.

Seen in relation to Marxism, Heidegger's final system seems to call for the comparison with Marx's and it is, indeed, surprising that so little has been done along these lines. The important influences of Heidegger on Marxism tend to be highly indirect: e.g., through the philosophical hermeneutics of Hans-Georg Gadamer and within the context of French structuralism. By contrast, the interpretation presented here aims at confronting Heidegger's mature thought head-on with a viable reading of Marx. The central themes will accordingly be: Heidegger's claim for a priority of Being, his doctrine of the forgetfulness of Being and the structure and methodology of his critical meta-ontology – especially the relation of its concepts to history. The basic analysis of Heidegger's system attempts to capture what seems to be obviously at work in Heidegger's writings since the mid-thirties in line with reflections which Heidegger himself makes in his latest work. The danger is, of course, that any such over-all sketch is reductive of Heidegger's thought, whose importance lies more in its concrete suggestions and specific points then in its general outlines – witness the above reference to hermeneutics and structuralism. If, however, this interpretation lacks the profundity that alone can benefit from Heidegger, at least it consciously avoids the shortcomings of previous interpretive attempts and clears the way for further work by establishing a context within which the confrontation between Marx and Heidegger can meaningfully be developed. Although placed within a critical argument, the interpretation of Heidegger, like that of Marx, aims at sympathetic understanding and constructive development.

The problem with previous interpretations of Marx (including Heidegger's) and of Heidegger (including those by Marxists) can be summed up in one objection: they impose a preconception upon their object. This is precisely what phenomenology, with its slogan: *"Zur Sache selbst,"* rebelled against. Heidegger has adopted this ethos in demanding that Being-itself be thought about "appropriately." Appropriate thought appropriates its object in an appropriate way, in a manner derived from the thing itself. Marxism, too, in line with its rejection of ideology, is opposed to criticism from an external standpoint; Marx's "immanent critiques" of Hegel, political economy and bourgeois ideology in general set out from the presuppositions of the questionable theory itself in order to show its contradictions and inadequacies.

To understand Marx and Heidegger appropriately, to uncover what is unique and original to each, means to follow their own hermeneutic principles. In comparing the two systems, neither can be subordinated to the other or to some supposedly objective third standpoint of commonsensical analysis. The principle guiding the present work has been to allow the two systems to unfold themselves autonomously, understanding the tangential points as organic parts of their respective contexts. This has been sought through keeping the two presentations distinct rather than comparing them point by point. The systems are developed through close textual analysis of key works, which, however, are selected with an eye to the comparison. Further, the confrontation is not externally imposed; it arises immanently out of the present crisis of Marxist theory and the contradictions of Heidegger's thought as well as out of the internal demands of the two systems. Once the Marx interpretation has been spelled out, the points of comparison can be developed in terms of the material as it occurs in the course of the Heidegger presentation, thereby strengthening the focus of the Heidegger interpretation without distorting it.

Just as, for Marx, immanent critique need not become apologetic if it retains its critical thrust, so, for Heidegger, what is decisive is not to avoid the hermeneutic circle but to come into it in the right way: "Our first, last and constant task is never to allow our fore-having, fore-sight and fore-conception to be presented to us by fancies and popular conceptions, but rather to make the scientific theme secure by working out these fore-structures in terms of the things themselves." [4]

As this quotation from Heidegger notes, it is not merely one's project and an anticipation of the results which form preconditions of understanding, but one's preconceptions as well. If one is to avoid external critique which is inappropriate, distorts and misses the point, then account must be taken of the source of preconceptions, the *Wirkungsgeschichte* of the work under consideration.[5] Only through the history of its effects, its tradition of having

---

[4] Martin Heidegger, *Being and Time* (New York: Harper & Row, 1962), p. 195. Cf. Martin Heidegger, *Sein und Zeit* (Tübingen: Niemeyer, 1967), S. 153.

[5] This notion of the importance of the historical effects of a text on the subsequent comprehension of that text is developed in Hans-Georg Gadamer, *Wahrheit und Methode: Grundzüge einer philosophischen Hermeneutik*

been variously construed, does a work cross the gap between the author and the reader. The history of ideas is thus the medium which permits understanding, the reconciliation of the dead spirit in language with that spirit which forces it to life on the basis of its afterlife.

But intellectual history takes place in the context of socio-political developments. Heideggerian hermeneutics may be correct when it argues that society can only be known through linguistic texts: "Language is the house of Being" and conversely "Being, which can be understood, is language."[6] Thus, it is true that Marx focuses on Hegel's texts, the tomes of bourgeois political economy and British governmental reports. More generally, "society" is to be located only in its citizens, that is, in their (fundamentally linguistic) objectifications in self-reflection, speech, documents, works and institutions. Marxism none-the-less has the last word when it points out that the subjects have already been thoroughly mediated, so that the social superstructure created by their activity is, through them, already (pre-linguistically) shaped by the character of the social totality. Karl-Otto Apel is thus right to point to the basis in the "community of interpreters" for the ontological categories, whose history Heidegger leaves to a linguistic world-spirit whose theological overtones have merely been modernized and whose substance has accordingly been diminished.[7] However, in abstracting from the historically-specific to formulate the ideal of a speech community, Apel is himself in danger of abstracting from the social context of the communicating subjects: their relations within a specific, concrete, historical form of production.

A merger is necessary between the hermeneutic insight into the context-dependence of all understanding and the ideology-critical emphasis on

---

(1960; 2nd ed. Tubingen: Mohr, 1965).

[6] These central motifs of Heidegger's thought are elaborated in Gadamer's discussion, especially in the Preface to the second edition of *Wahrheit und Methode*.

[7] Cf. Karl-Otto Apel, *Transformation der Philosophie*, 2 vols. (Frankfurt: Suhrkamp, 1973), especially the extensive Introduction to the first volume. This introduction represents the latest stage in the debate between hermeneutics and ideology critique, demonstrating Apel's role as innovative interpreter of both Heidegger and Marxism. The dispute, the most extended and explicit confrontation of the thought of Marx and Heidegger to date, began with Habermas' critique of Gadamer's *Wahrheit und Methode* in the former's "Zur Logik der Sozialwissenschaften" (*Philosophische Rundschau*, Beiheft 5, February 1967). Subsequent contributions to the debate have been collected in *Continuum* (vol. 8, nos. 1&2, 1970) and *Hermeneutik und Ideologiekritik* (Frankfurt: Suhrkamp, 1971).

societal mediations. From his early analysis of Being-in-the-world as the essential structure of human existence, Heidegger has stressed the importance of the world around a being to the character of the being itself: a tool has meaning within a technical context, a jug within the relationships of the physical world, a bridge within lived space and a word within the communicative situation. The grand question of Being is ultimately an investigation of the contextuality of beings. But Heidegger fails to recognize the power of social formations to define the context of beings; here Marxism furnishes the antidote. With Marx, *social theory supplies the comprehension of the context.* A Marxist appropriation of Heidegger's critique of non-contextual, "metaphysical" positivism would simultaneously clarify Marxism's own approach and demystify Heidegger's content-poor ontological musings. For Marx and his creative followers have articulated numerous ways in which the power of the context to structure the beings it contains is itself created by those beings. Such analyses are, however, readily subject to misunderstandings unless they are formulated within a theory which explicitly rejects mechanistic, positivistic, idealistic and subjectivistic philosophical stances. To bring out those fundamental theoretical features of Marx's thought which are especially important today requires a peculiarly twentieth century formulation which would make explicit how social facts are comprehended within a social theory and how the categories and orientation of that theory are related to its social context.

Because the essence of man inheres in the nexus of social relations from the viewpoint of social theory, human activity constitutes social *praxis*, the process of the production and reproduction of the form and substance of society. The task of socially-conscious theory is accordingly to interpret social phenomena, as human artifacts and, as such, as the expression of social relations among people. The reconstruction of the preconditions of the given social reality should ideally demonstrate the mediations that constitute its history. This demonstration is neither a recounting of empirical history, a logical argument unrelated to the specifics of the case, nor a causal account of events and effects. Rather, it points to ways in which the phenomena have been conditioned, have been characterized by social conditions such that in the end the social origins have become obscured. Political events, for instance, function as both symptoms and screens for social transformations.

Outside the political arena of the past century, divorced from the Russian and Chinese revolutions, the failure of revolution in the West, the rise of fascism, the development of advanced industrial economies and culture, Marx cannot today be understood. For it is in terms of such events and what underlies them that the Leninist, Stalinist, Maoist, existentialist and humanist interpretations arose.

A contemporary understanding of Marx must take into account these events, the social relations behind them and the resultant interpretations if it is to comprehend its own procedure, possibilities and necessity. The situation with respect to understanding Heidegger is, if slightly less complex, not as different as might be assumed. What is particularly clear in *Capital* holds for Heidegger's writings and his references to Marx as well: namely, the philosophical argument is inextricable from timely observations and social considerations. This relates as much to the perspective of the reader as to that of the authors.

It is precisely our temporal distance from the concerns of the past decades which makes the following interpretations possible. Until recently, the hermeneutic goal of understanding the author better than he understood himself has been hindered by politics. In their concern to battle socialism and Stalinism, the Heideggerians ignored or distorted the thought of the man the Soviets claimed as a founding father. Similarly, Marxists felt compelled to attack and ridicule the thought of a philosopher who had consorted with fascism, and here the "existentialist" themes seemed most vulnerable. This is not to imply that the problems underlying the old politics and polemics have vanished nor that exegesis must or can completely disavow politics. But philosophy today is in a period of retrenchment, where hasty formulations prove ineffectual; serious interpretation of Marx and Heidegger is presently underway throughout the world. This has opened the possibility of a successful confrontation of their respective systems, already implicit in the convergence of approaches and concerns in the respective philosophical camps. The political changes are, of course, related to social conditions which are more difficult to evaluate. Suffice it to say that developments in the consciousness of youth throughout the world in the past decade suggest progress in the conditions of the possibility of a new epoch in both Marx's and Heidegger's terms. If this is so, then the Marxian and Heideggerian systems have gained in relevance, and that means in accessibility and comparability.

The point of new interpretations of Marx and Heidegger is not to rewrite *Capital* and *Being and Time* as though *sub specie eternitatis;* rather, each age – every decade, class and country – requires its own understanding, incorporating both changes in the social fabric and consequent modifications in revolutionary perspectives. That the American New Left considered *Capital* irrelevant is understandable; whatever unfortunate consequences it may have had, this attitude allowed for a freshness, creativity and experimentation which may not only have been its greatest virtue, but its only objective potential. The 1970's, however, call for a synthesis of the best in the old and new leftovers. The following is not the required reformulation of Marx and Heidegger, but understands itself as a faltering step in the task of clarification, analysis and interpretation which recognizes itself to be politically, historically and philosophically situated. This means that perspectives which may well be appropriate in Eastern Europe, Italy or China are here rejected. Not unrelated to the concern with the situation of advanced industrial society, the insights of Theodor W. Adorno have guided the whole of the dissertation. Acknowledgment is made therefore by quoting Walter Benjamin, Adorno's guru, whose ephemeral and contradictory character may provide an appropriate symbol for the iconoclastic attitude of the so-called Frankfurt School.

In line with their tentative character, the following presentations can be taken as theses on reading Marx and Heidegger today, working hypotheses for future inquiry. Accordingly, the thought of Marx and Heidegger, which is conceived of as systematic, as well as the debates between them are presented in terms of their development. Rather than starting with texts which represent mature statements of the systems, the analyses unfold in chronological order, even if the continuity and teleology of thought is often stressed over the deviations. Not the least motivation for this procedure is the suspicion that the System has become an anachronism. Where systematic presentations tend to petrify into monuments, an approach which follows the research which spirals in on a system makes more sense pedagogically and critically, for it stresses the arguments and *aporia*. Nevertheless, the mature works of Marx and Heidegger assume a priority in the interpretation of their early works, which are grasped as the seeds of the later thought and thus as inadequate articulations of that which they intend.

Chapter II. Heidegger's Critique of Marx

Being and Time (1927) made its appearance between two of the most important Marxist publications since *Capital* (1867), namely Georg Lukacs' *History and Class Consciousness* (1923) and Marx's 1844 *Economic-Philosophic Manuscripts* (first published in 1932). It is perhaps less arbitrary to place Heidegger in this context than might be assumed, Heidegger formulates both the historical motivation for the analyses in *Being and Time* and the task which remains at its end in terms of the concept of "reification of consciousness," the category central to the philosophically most important essay in Lukacs' book and later made popular by the discussion of "alienation" in Marx's *Manuscripts*.

Although Heidegger never wrote extensively on Marx or explicitly referred to Lukacs, those references to Marx which he does make assign him a surprisingly central position within the field of Heidegger's concerns, Significantly, the dozen references to Marx's thought which occur in Heidegger's later writings deal exclusively with Marx's early manuscript and the *Theses on Feuerbach* of a year later. A review of Heidegger's comments on Marx in terms of their misunderstandings as well as their insights raises the suspicion that Heidegger's familiarity with Marx is limited to a superficial reading of these early writings, an attempt to dismiss Lukacs' philosophy as insufficiently radical (in the philosophical sense), a sympathy for conservative social criticism and even an openness to propagandistic anti-communism. This impression is, of course at odds with Heidegger's carefully cultured reputation as a thorough historian of philosophy, a discoverer of what has remained implicit in what is said and a thinker of Being whose inspirations are above merely empirical, political influences.

In view of the importance Heidegger quietly attributes to Marx's thought, one is forced to wonder why he never dealt with Marx in anything like the way he delved into Nietzsche. The suspicion that this represents an important failing in Heidegger's work is a central motivation of the present comparison of Marx and Heidegger. In order to orient this study on the central issues and to incorporate what thinking Heidegger has devoted to the question of his relation to Marx, it is useful to consider hints Heidegger has, almost parenthetically, sprinkled through his writings. This chapter shall, therefore, review all his published references to Marx, Marxism and materialism.

Heidegger's Early Criticisms

To begin with, it is interesting to note what *Being and Time* had to say about the Lukacsian phrase, "reification of consciousness." There are three passages to consider. At the very start of *Being and Time*, where Heidegger is motivating the investigation of his fundamental ontology of *Dasein* (human Being), he argues that even the analysis of reified consciousness, no matter how critical it may be of the present human condition, still assumes uncritically the traditional concept of subjectivity as a standard:

> *Ontologically*, every idea of a 'subject' – unless refined by a
> previous ontological determination of its basic character –

> still posits the *subjectum* (*hypokeimenon*) along with it, no matter how vigorous one's ontical protestations against the 'soul substance' or the 'reification of consciousness.' Thinghood itself which such reification implies must have its ontological origin demonstrated if we are to be in a position to ask what we are to understand *positively* when we think of the nonreified *Being* of the subject, the soul, the consciousness, the spirit, the person.[8]

Where Lukacs' analyses reified consciousness from within a problematized subject/object metaphysics (unlike Marx, as Heidegger fails to see), Heidegger sets out from the phenomenon of reified consciousness (formalized in the new terminology as inauthentic *Dasein*) in order to get at the essence behind this appearance. By raising the transcendental question of the conditions of the possibility of reification or inauthenticity, Heidegger hopes to arrive at a non-dogmatic conception of authenticity. The dialectic between the abstract value of capitalist commodities and their concrete use value, which is at the base of Lukacs' analysis of reification, is cast in the aura of a radical ontological investigation in terms of presence-at-hand and readiness-to-hand. Heidegger's originality here lies in his relating ontological categories to temporal structures – to human temporality in *Being and Time*.

Toward the end of his major work, Heidegger indicates that his analysis of temporality is intended to show the superiority of his analysis of presence-at-hand over Lukacs' treatment of reification:

> If world-time thus belongs to the temporalizing of temporality, then it can neither be violated 'subjectivistically' nor 'reified' by a vicious 'objectification'.[9]

If the last part of this sentence is, indeed, aimed at Lukacs, it does the depth of Lukacs' analysis little justice. Lukacs in fact gave a coherent argument to show how human "temporalizing" was historically transformed into "world-time" due to social changes related to the transformation from the feudal to the capitalist mode of production. Lukacs' eminently Marxian analysis suggests a mediating link between changes in ontological categories and societal developments, precisely the type of connection that is missing in Heidegger's entire path of thought. Further Lukacs quotes Marx as having in 1847 (in *The Poverty of Philosophy*) already noted the reification of qualitative

---

[8] Martin Heidegger, *Being and Time*, p. 72, S. 46.

[9] Ibid., p. 472, S. 420. Cf. Georg Lukacs, *Geschichte und Klassenbewusstsein* (Berlin: Malik, 1967), S. 100f.

temporality into the quantitative measurement of time as a consequence of the mechanization of production.

In the end, it is unclear just how Heidegger's analysis of the relationship between reification and temporality is supposed to be superior to Lukacs'. On the final page of *Being and Time*, Heidegger calls his accomplishments merely the "point of departure" and indicates that all of the crucial questions about reification remain to be settled:

> The distinction between the Being of existing *Dasein* and the Being of being which does not have the character of *Dasein* may appear very illuminating: but it is still only the *point of departure* for the ontological problematic; it is nothing with which philosophy may tranquilize itself. It has long been known that ancient ontology works with 'thing-concepts' and that there is a danger of 'reifying consciousness.' But what does this reifying signify? Where does it arise? Why does Being get 'conceived' 'proximally' in terms of the present-at-hand *and not* in terms of the ready-to-hand, which indeed lies *even closer*? Why does this reifying always keep coming back to power? How is the Being of 'consciousness' *positively* structured such that reification remains inappropriate to it? Is the 'distinction' between 'consciousness' and 'thing' sufficient for tackling the ontological problematic in a primordial manner? Do the answers to these questions lie along our way? And can the answer even be *sought* as long as the *question* of the meaning of Being remains unformulated and unclarified?[10]

There is an ambiguity to Heidegger's relationship to Lukacs, which foreshadows his later attitude to Marx. It is not clear whether Heidegger – who claims his analysis is more fundamental than Marxism – wishes to reject the thought of Lukacs and Marx or unobtrusively to translate it into a new conceptualization. While the question is basically a matter of degree, the opposed strivings do both seem to be at work in Heidegger's writings. Whatever the intention of Heidegger's references to Lukacs, they clearly present two characteristics of his approach which are opposed to Marxism and which Adorno singles out for criticism: the attempt to push all questions back to a *fundamental* question and the search for *positive* structures to replace Marxism's negative, but therefore critical, analyses.

---

[10] Ibid., p. 487, S. 436f.

Political Distortions

In the context of Heidegger's life's work, *Being and Time* represents the starting point of his research, of his path of thought. But much of its approach was later rejected. During the 1930's Heidegger reversed his opinion concerning the *"Sache des Denkens,"* the essential theoretical question which was to stand at the head of his system. From a concern with the temporality of the *individual,* Heidegger turned to a meditation on that which assures *philosophy* a possible history, i.e., on the conditions of the possibility of an epochal (historical) structure to the ontological presuppositions that characterize the presence of beings. The first major public presentation of this new problematic was Heidegger's *Letter on Humanism*, which took the occasion of a disagreement with Sartre to unfold Heidegger's own position. Considering the extensive and important discussion of Marx in this essay, it can well be considered an attempt to present Heidegger's thought as an alternative to Marxism rather than to existentialism.

Gajo Petrovic, who analyzes Heidegger's comments on Marx in his article on "Der Spruch des Heideggers," argues that Heidegger has merely indicated a basis for discussion between the two viewpoints, but has declined to proceed to the comparison.[11] Thus, Heidegger has pointed to two aspects of Marx's thought which make it important to the Heideggerian project: the concept of alienation and the recognition of the historical in Being. However, according to Petrovic, Heidegger has failed in his understanding of Marx to recognize the unity of the latter's thought and, relatedly, to comprehend it in its full originality. In assuming that Marx's approach is metaphysically humanistic, Heidegger consistently misinterprets it, fitting it neatly into the history of metaphysics without considering what is unique to Marx, and consequently failing to learn from him or even to join in a fruitful conversation with him. To correct the shortcomings of Heidegger's attempt at comparing his own thought with Marx's requires an interpretation of the core of each system in terms of its respective originality. This has been attempted in the following parts of the present work. The results reached there are here anticipated in order to review and evaluate Heidegger's understanding of Marx.

It is important first of all to note the developmental character of Heidegger's conscious, or at least published, relationship to Marx. *Being and Time* and the other early writings never mention Marxism despite its extraordinary significance in the intellectual atmosphere of a seemingly pre-revolutionary Germany – at most a facile dig is made at Lukacs. In the war years, when

---

[11] Gajo Petrovik, "Der Spruch des Heideggers," *Durchblicke. Martin Heidegger zum 80. Geburtstag* (Frankfurt: Klosterman, 1970), S. 412ff.

Heidegger was meditating upon spirit, art and Nietzsche in an attempt to rebut Nazi ideology, he identified Marxism with the mechanistic simplifications of crude Diamat (Marxism-Leninism), rather than trying to develop a humane and critical Marxism as did other independent thinkers – Adorno and Merleau-Ponty, for instance. For Heidegger, Marxism is viewed as just as much of the social problem as fascism:

> The spirit falsified into intelligence thus falls to the role of a tool in the service of others, a tool the manipulation of which can be taught and learned. Whether this use of intelligence relates to the regulation and domination of the material conditions of production (as in Marxism) or in general to the intelligent ordering and explanation of everything that is present and already posited at any time (as in positivism), or whether it is applied to the organization and regulation of the mass and race of a folk, in any case the spirit as intelligence becomes the impotent superstructure of something else, which, because it is without spirit or even opposed to the spirit, is taken for the authentic reality. If the spirit is taken as intelligence, as is done in the most extreme form of (by?) Marxism, . . . (1935)[12]

There is no attempt made here to distinguish what is of value in Marx's thought from its vulgar distortion. Around 1940, when he was engaged in a monumental task of interpreting Nietzsche in explicit opposition to the prevailing interpretation by the Nazis, Heidegger still seems to have uncritically accepted the Nazi view of Marx as a political ideologue with no philosophical originality to offer.

> That the medieval theologians study Plato and Aristotle in their own way, i.e., giving them a new meaning, is the same as that Karl Marx used Hegel's metaphysics for his own political world-view. (1940)[13]

Not until the *Letter on Humanism* is Marx taken as a serious thinker.

> But whence and how is the essence of man determined? Marx demands that the 'human man' be known and acknowledged. He finds this man in 'society'. The 'social' man is for him the 'natural' man. In 'society' the 'nature' of

---

[12] Martin Heidegger, *An Introduction to Metaphysics* (Garden City: Anchor, 1961), p. 38f. Cf. Martin Heidegger, *Einführung in die Metaphysik* (Tübingen: Niemeyer, 1958), S. 35f.

[13] Martin Heidegger, *Nietzsche* (Pfullingen: Neske, 1961), Bd. II, S. 132.

man, which means all of his 'natural needs' (food, clothing, reproduction, economic sufficiency), is equally secured. (1946)[14]

Even here, Heidegger's pronouncements are problematic, as Petrovic points out. Heidegger puts Marx's key terms in quotation marks to indicate that they are questionable without bothering to question what Marx meant by them. There is no recognition on Heidegger's part that Marx uses the adjectives "human," "social" or "natural" in a critical way: as dialectically opposed to "alienated." On the contrary, Heidegger implies that the use of these terms makes Marx into a traditional, metaphysical humanist, a writer who merely accepts the dogmatic view of humanity as having a fixed essence. That Marx developed his concepts through a *critique* of Hegelian metaphysics – a specific negation, not a simple inversion – suggests that Marx may have escaped the metaphysical position, particularly in his mature works where the Hegelian terms rarely appear even in their critical form. Further, the concern with securing economic sufficiency is so reductive of Marx's thought that it is more appropriate to that non-Marxian, non-philosophical "materialism" Heidegger refers to elsewhere.[15] But such "materialism," the greed for material goods as opposed to the higher "values," is unrelated to Marx except as ignorant caricature.

The next mention of Marx puts him in good company in Heidegger's scenario: right along with Nietzsche:

> Absolute metaphysics belongs with its inversions by Marx and Nietzsche to the history of the truth of Being. (1946)[16]

This evaluation of Marx is repeated in later years without further explanation:

> But in what does the *telos* consist, the consummation of modern philosophy, if we may speak of such? In Hegel or not until Schelling's late philosophy? And how about Marx and Nietzsche? (1955)[17]

And again:

---

[14] Martin Heidegger, "Letter on humanism," *Philosophy in the Twentieth Century* (New York: Harper & Row, 1971), vol. III, p. 197. Cf. Martin Heidegger, "Humanismusbrief," *Wegmarken* (Frankfurt: Klostermann, 1967), S. 151.

[15] Cf. Martin Heidegger, *Der Satz von Grund* (Pfullingen: Neske, 1971), S. 199f.

[16] "Letter on humanism," p. 206, S. 166.

[17] Martin Heidegger, *What is Philosophy?* (bilingual ed., New Haven: College & University Press, n.d.).

> Marx and Nietzsche are the greatest Hegelians. . . . The consummation *is* only as the total process of the history of philosophy, in which process the beginning remains as essential as the consummation: Hegel and the Greeks. (1958)[18]

While Heidegger spent several years and over a thousand published pages to explain how Nietzsche had inverted the metaphysics which held sway from Plato's *Republic* to Hegel's *Logic* and Schelling's essay on human freedom, he has not dedicated a single phrase to the possibility that Marx's *Capital* might have left that tradition behind – as Heidegger's own *Verwindung* followed his *Überwindung* of metaphysics. Having struggled so hard to learn from the example of Nietzsche's failure to transcend metaphysics, Heidegger seems to have avoided raising the question whether Marx might have something positive to contribute. It seems that Heidegger's political conservatism and his flirtations with an existentialist jargon led him to ignore Marx in favor of Nietzsche until his own thought had really developed and he could see the parallels with Marx. Not only would an earlier study of Marx have saved Heidegger from traveling down some dark dead-end trails ("*Holzwege,*" as he calls them), but a more profound understanding of Marx might still help to fill in some content in the emptiness of Heidegger's concepts.

Heidegger suggests that a productive discussion with Marxism would focus on the related terms "alienation" and "homelessness," both understood in relation to an essential dimension of history – not psychologically. The key to the analysis would be a consideration of the essence of technology. Through an understanding of the essence of technology, one could discover why today everything appears as the material of labor.

> Homelessness becomes a world destiny. It is, therefore, necessary to think of this destiny from the point of view of the history of Being. What Marx, deriving from Hegel, recognized in an essential and significant sense as the alienation of man, reaches roots back into the homelessness of modern man. This is evoked – from the destiny or Being – in the form of metaphysics, strengthened by it and at the same time covered by it in its character as homelessness. Because Marx, in discovering this alienation, reaches into an essential dimension of history, the Marxist view of history excels all other history. Because, however, neither Husserl nor, as far as I can see, Sartre recognizes the essentially

---

[18] Martin Heidegger, Hegel und die Griechen," *Wegmarken* (Frankfurt: Klostermann, 1967), S. 260f.

historical character of Being, neither phenomenology nor existentialism can penetrate that dimension within which a productive discussion with Marxism is alone possible.

For this it is necessary to liberate oneself from the naive conceptions of materialism and from the cheap, supposedly effective, refutations of it. The essence of materialism does not consist of the assertion that everything is merely matter, but rather of a metaphysical determination according to which all beings appear as the material of labor. The modern metaphysical essence of labor is anticipated in Hegel's *Phenomenology of Spirit* as the self-establishing process of unconditional production; i.e., the objectification of the actual through man experienced as subjectivity. The essence of materialism is concealed in the essence of technology, about which, indeed, a great deal is written, but little is thought. Technology in its essence is a destiny (in the history of Being) of the truth of Being resting in forgetfulness. (1946)[19]

Until one associates the term "materialism" with Marx as a dialectical materialist, what Heidegger says is fine. But Heidegger clearly does make this identification, thinking he is saving Marx's philosophy from interpretations that are even more naive.

It is crucial to show that Marx does not simply posit all beings as material of labor and nothing more, for it is this supposed assumption which makes Marx a metaphysical thinker in Heidegger's eyes, relating him to both Hegel and Nietzsche. This metaphysical position is appropriately attributed to Ernst Junger's non-Marxian book, *Der Arbeiter*, which Heidegger carefully studied, but not to Marx's writings.[20] Even within the labor process, Marx

---

[19] "Letter on humanism," p. 209, S. 170.

[20] Ernst Jünger, *Der Arbeiter: Herrschaft und Gestalt* (Hamburg, 1932). Heidegger praises this book for having "achieved a description of European nihilism in its phase after World War I" and for making "the 'total work character' of all reality visible from the figure of the worker." He characterizes it as "a clear-sighted book" which understandably "was being watched and was finally forbidden" by the Nazis. Cf. Martin Heidegger, *The Question of Being* (bilingual ed., New Haven: College & University Press, 1958), p. 40ff. Herbert Marcuse (Herbert Marcuse, "The affirmative character of culture," *Negations*. Boston: Beacon Press, 1969, p. 128) presents a different evaluation of Jünger's book, viewing it as part of the ideological preparations for German fascism in the 1930's:

distinguishes between that moment of an object which is mere grist for the mills of capitalist industry and that indissoluble kernel of nature which is in principle out of humanity's reach.[21] In *Capital* it is only as abstract exchange value, not as concrete use value, that the commodity corresponds to what Heidegger has in mind as material of labor. Further, abstract value does not represent a metaphysical assumption on the part of Marx; he understands it as an historical appearance, a fetish, which reaches its extreme form in automation. Marx's concept remains critical of the present metaphysical character of beings as commodities in terms of both a pre-capitalist past and a projected future. Marx's epochs and their political content may not correspond exactly to Heidegger's scheme, but surely the two share a critical distance from the contemporary "metaphysical determination" which Heidegger attributes to the essence of materialism. The reduction of nature to material for labor is, in Marx's account, a social product, not the result of his or Hegel's autonomous speculation. Marx's "materialism" consists precisely in the thesis of the primacy of societal mediations in determining ontological categories: the very principle Heidegger repeatedly overlooks in his interpretation of Marx and consistently lacks in his own thought.

In later comments, Heidegger rather flippantly casts Marx aside, but only to underline the crucial question underlying his attempt to provide an alternative to Marxism: What is the heart of the matter, the essence which must determine the path for critical thought today? In the process, Heidegger suggests that Marx failed through too great an eagerness for action:

> None of us know what craft modern man must engage in in the technical world, must engage in even if he is not a worker in the sense of a worker on the machine. Even Hegel, even Marx could not yet know this and ask this,

The cynical suggestions offered by Jünger are vague and restricted primarily to art. 'Just as the victor writes history, i.e., creates his myths, so he decides what is to count as art.' Even art must enter the service of national defense and of labor and military discipline. (Jünger mentions city planning: the dismemberment of large city blocks in order to disperse the masses in the event of war and revolution, the military organization of the countryside, and so forth.) Insofar as such culture aims at the enrichment, beautification and security of the authoritarian state, it is marked by its social function of organizing the whole society in the interest of a few economically powerful groups and the hangers on.

[21] The indissoluble kernel has been likened to the Kantian *Ding-an-sich* — transformed from the realm of conception to that of labor — by Alfred Schmidt in his *Marx's Concept of Nature* (London: New Left Books, 1972).

because even their thought still had to move in the shadow of the essence of technology, which is why they never freed themselves to think about this essence adequately. No matter how important the socio-economic, the political, the moral, even the religious questions may be, which are handled in relation to the technical craft, none of them reach at all in the yet unthought essence of the manner in which everything *is* at all which stands under the domination of the essence of technology. This has remained unthought because the will to act, i.e., to make and cause, has crushed thought. (1952)[22]

Further:

> However, the transformation of the world so considered requires first that thought change, just as behind this (Marx's) demand a transformation of thought already stands. (Cf. Marx, *Theses on Feuerbach*, 11) But in what manner should thought change itself if it does not set itself on the path into that which is worthy of thought? (1962)[23]

Heidegger is justly well-known for his quietism, his contemplative stance of letting Being be. Insofar as this is not merely a way of making a methodological demand similar to that of descriptive phenomenology, it may be understandable as a political judgment in times very different from Marx's. After all, on the one hand Heidegger enthusiastically jumped on the bandwagon of activism in the early 1930's and it is understandable that he would seek to avoid that error being repeated. On the other hand, the example of Adorno shows that even a Marxist may have to argue against prevalent forms of activism and resign himself to contemplation as a result of the political climate.

Marx's point – according to his eleventh thesis which Heidegger insists on not understanding despite his adherence to it – is that philosophical interpretation must not be an end in itself, but must be part of a hermeneutic circle which includes the anticipation of a transformed world, thereby intervening critically in the given reality. However, there is something to Heidegger's argument. Marx's activism is related to his view of the centrality

---

[22] Martin Heidegger, *What is Called Thinking?* (New York: Harper & Row, 1968), p. 24.

[23] Martin Heidegger, "Kant's These über das Sein," *Wegmarken* (Frankfurt: Klostermann, 1967), p. 274f.

of the socio-economic, the political questions, answers to which demand conscious human action.

Heidegger's alternative, that the essential question is the question of Being, does not so obviously involve political action, but seems to belong more to the contemplative realm of the study of philosophy. Nevertheless, Heidegger opens the *Letter on Humanism* with the statement that, "Our thinking about the essence of action is still far from resolute enough," suggesting that thinking about Being may also be a form of social action. For Marxists, it is clear that ontological reflection is not divorced from social practice, for better or worse. Again, the problem of activism and quietism in Marxism and Heidegger may be helped to its resolution through the discussion between them.

Heidegger's Mature Criticisms

In the mature statement of his system, the essay *Time and Being* (1962), Heidegger does not list Marx's "metaphysics" next to Nietzsche's in the history of philosophy; he entirely ignores Marx. However, in subsequent reflections on the relation of his own thought to other recent philosophers or to the task of thought today, Heidegger is still very much concerned with Marx. In the following passage he still views Marx as metaphysically grounding the way things are in the "dialectical mediation of the movement of the historical process of production."

> Metaphysics thinks about beings in the manner of representational thinking which grounds (with reasons). For since the beginning of philosophy and, with that beginning, the Being of beings has shown itself as the ground (*arche*, *aition*, principle). The ground is that from which beings as such are what they are and how they are in their becoming, perishing and persisting as knowable, manipulated, worked. As the ground, Being brings beings to their respective presencing. The ground shows itself as presence. The present of presence consists in the fact that it brings what is present each in its own way to presence. In accordance with the kind of presence, the ground has the character of grounding as the ontic causation of the real, as the transcendental making possible of the objectivity of objects, as the dialectical mediation of the movement of the absolute spirit, of the historical process of production, as the will to

> power positing values. What characterizes metaphysical
> thinking, which grounds the ground for beings, is that
> metaphysical thinking, starting from what is present,
> represents this in its presence and thus presents it in terms
> of its ground, as something grounded. (1964)[24]

There is something to this: it recognizes the historical dimension of Marx's analysis and it is true that for Marx the character of beings is dependent upon the predominant conditions of production in society. Further, Marx sometimes suggests that all history can be viewed in these terms – this outlook has subsequently been systematized under the name historical materialism and turned into a truly metaphysical dogma.

However, Marx can at least also be interpreted as arguing that the primacy of commodity production – as the *essential* category of social analysis, rather than simply as one *precondition* of social existence – is itself historically-specific; that this primacy is part of the problem with capitalism; and that this primacy must itself be explained, i.e., taken as a symptom. This latter view of Marx is part of a more sophisticated understanding of his methodology, one which places him in close proximity to Heidegger. When incorporated into the comparison of Marx and Heidegger, this interpretation not only speaks to the inherent needs of Heideggerian theory, but benefits from the confrontation itself in terms of problematizing its foundations. For, as the last quotations suggest, Heidegger's main claim is that he is being philosophically more radical (*ursprünglich*) than Marx. A Marxist can argue that it is not the "final questions" which are important – even assuming they make sense or can be answered – but those more modest questions which are abstract enough to make possible a critical theory, but specific enough to be useful. Both Marx and Adorno, for instance, make this reasonable point. Yet, Heidegger's challenge here is more specific: does Marx's thought "represent" beings as "grounded" in a way which Heidegger avoids?

Heidegger's claim rests upon the assumption that Marx simply reversed traditional philosophy, retaining its unfortunate habit of grounding all beings in some particular, higher being:

> Throughout the whole history of philosophy, Plato's
> thinking remains decisive in changing forms. Metaphysics is
> Platonism. Nietzsche characterizes his philosophy as
> reversed Platonism. With the reversal of metaphysics, which

---

[24] Martin Heidegger, "The end of philosophy and the task of thinking," *On Time and Being* (New York: Harper & Row, 1972), p. 56. Cf. Martin Heidegger, "Das Ende der Philosophie und die Aufgabe des Denkens," *Zur Sache des Denkens* (Tübingen: Niemeyer, 1969), S. 62.

was already accomplished by Karl Marx, the most extreme possibility of philosophy is attained. It has entered its end. (1964)[25]

Heidegger would have us believe that Marx transformed representing, grounding metaphysics into empirical science (political economy, sociology, political science, anthropology), which grounds everything in a preconceived notion of its object: man or society. We have already indicated that this is a distorted view of Marx's dialectical, critical, emancipatory "science." Even if Heidegger can get away with claiming that "normal" science (research within a paradigm) does not think, i.e., does not question its foundations, that cannot be extended to Marx, no matter how often Marx used the term "science" or how much content his concepts articulate, for he rejects the need for foundations *a la* fundamental ontology. Calculative thinking adds richness to Marx's thought, for it informs his conceptual framework rather than presupposing it; his empirical research is dialectically intertwined with its own guiding theory as expressed in the systematic presentation.

Heidegger's final reference to Marx, in a 1969 television broadcast, indicates three further criticisms:

(1) Talk of society is metaphysical because society is posited as an absolute (unconditioned) subject (agent).

(2) Marx is involved in "representing" the world, which, according to Heidegger's essay on the "Age of the World-view," involves grounding the world in the interpreting subject.

(3) Marxism remains philosophically within the subject/object relation and therefore cannot grasp the essence of technology.

> (Professor Heidegger, . . . Do you see a social mandate for philosophy?)

> No. One cannot speak of a social mandate in this sense. If one wants to answer this question, one must first ask: "What is society?" and must then reflect that today's society is just the absolutizing of modern subjectivity and that from this perspective a philosophy which has overcome the standpoint of subjectivity cannot enter the discussion. Another question is, to what extent one can speak of a *transformation* of society. The question about the demand to transform the world leads to a much-quoted sentence by Karl Marx in the *Theses on Feuerbach*. I would like to cite it

---

[25] Ibid., p. 57, S. 63.

exactly and read it: "The philosophers have merely *interpreted* the world differently; the important thing is to *transform* it." In citing this sentence *and* in following this sentence, one ignores the fact that a transformation of the world presupposes a change of the *representation of the world* and that a representation of the world can only be won when one *interprets* the world sufficiently.

That is, Marx bases himself on a certain interpretation of the world in order to demand its "transformation," and thereby we can see that this sentence is unfounded. It gives the impression of speaking decisively against philosophy, although in the second part of the sentence the demand for philosophy is silently presupposed. . . . I see, however, in the essence of technology the first glimpse of a much deeper secret, which I name the *Ereignis*, the event of appropriation – from which you can gather that there can be no question of a resistance to or a negative judgment of technology. Rather, it is important to understand the *essence* of technology and the technical world. It seems to me that that cannot occur as long as one remains philosophically within the subject/object relation. That is, the *essence* of technology cannot be understood on the basis of Marxism. (1969)[26]

(1) Heidegger implies that Marx conceived of society as the collective will of free subjectivities who had but to agree upon change for it to be accomplished. While Heidegger may have attributed this view to Marx on the basis of reading Lukacs, it is an unacceptable interpretation. For Marx, society is not a being, capitalism is not a thing; Marx's analysis aims precisely at dispelling such fetishisms. This naive view of society as absolute subject seems much more to underlie Heidegger's own enthusiasm for the Hitler state's act of taking its destiny into its own hands, as expressed in Heidegger's 1933 *Rektoratsrede*. It is precisely such a voluntaristic conception of society that Marx attacked in his arguments with liberalism, utopianism, anarchism and vulgar socialism. Far from calling for an arbitrary transformation of society, which would create a new social formation *ex nihilo* or by subjective will power, Marx developed a theory incorporating a strong sense of social destiny. Revolutionary freedom consisted, for him, primarily in the recognition and comprehension of the pervasive power of prevailing social relations and productive forces to define potentials and limitations within society and to condition any attempt at social transformations or conservations. The task of the revolutionary subject is thus given by his

---

[26] *Martin Heidegger im Gespräch* (Freiburg: Alber, 1969), S. 68f, 73f.

objective contexts to encourage the existing liberatory tendencies and possibilities while resisting the forces of reaction and repression. Perhaps the charge of absolute subjectivity is more plausibly directed against the process of production in Marx's account than against his concept of society as such. The outlines of the development of the modes of production – Asiatic, classical, feudal, capitalist, socialist – may give the appearance of an autonomous process of self-negation on the Hegelian model. However, the historical details in any of Marx's extended presentations – i.e., in *The German Ideology*, "Forms which Precede Capitalist Production" and *Capital*, versus the popular summaries in the "Preface" to *Towards a Critique of Political Economy* or in the *Communist Manifesto* – stress the interplay of the various kinds of objective conditions, geography, trade, politics, economic determinants, cultural biases, etc. Further, where the mode of production develops in conjunction with another factor – the establishment of a monetary system or the growth of scientific knowledge, for instance – neither is simply founded in the other; rather, Marx shows how they support each other dialectically as mutual preconditions.

(2) Heidegger's claim that Marx's representation of the world grounds the world in the interpreting subject is questionable on two counts. It has already been suggested that Marx's method consists in an interplay between research and systematic presentation. That means that the resulting representation of the world is derived from the reality which it "articulates" in the double sense of structuring and verbalizing. Secondly, for Marx, theories of society – including those ideologies which form the object of much of his research, as well as his own writings – cannot be divorced from the society they mirror. Thus even if Marx's representation of the world were shown to be grounded in his subjectivity, this is itself mediated through and through by objectivity and knows itself to be.

(3) Also problematic is Heidegger's condemnation of Marx's understanding of technology as remaining philosophically within the subject/object relation. It is by no means clear that one can ignore the subject/object dichotomy as simply as Heidegger has attempted. Hegel had already taken the alternative approach in trying to reconcile subject and object in the historical process of their dialectical development. Marx criticized Hegel's result as idealistic, arguing that the dichotomy had a basis in social reality and could therefore only be resolved through a transformation of the form of social relations. In the case of the subject/object relation, as in that of the essence/appearance distinction and the view of nature as material for labor, the factors in Marx's system are not dogmatic postulates to be discarded lightly, but aspects of reality under the constraints of the capitalist system. When Marx refers to the subject/object relation in his investigation of technology, it is not as a metaphysical principle of his system, but a part of the ideology that he is

subjecting to immanent critique. Marx's own theory of technology is based on his theory of surplus value, which is not directly related to a subject/object problematic.

This is where Heidegger's published position on Marx stands at the present and where it is likely to remain standing as far as Heidegger is personally concerned. These few explicit references are, of course, merely the surface appearance of the relation of the content of Heidegger's system to that of Marx. The task of interpretation is to bridge the gap between the explicit and the implicit. The ambiguity of Heidegger's style, which surrounds a poverty of apparent content with an aura of hidden profundity, makes this task slippery. The range of possibilities is wide. Has Heidegger fallen so far behind Hegel philosophically that he cannot comprehend Marx's advances? Or has his thought gone so far beyond us that it remains unintelligible? The truth of the matter probably lies near the center of the middle ground between these extremes; that, at least, is the heuristic principle of the present work.

Chapter III. A Marxist Critique of Heidegger

The jargon of authenticity is a social disease and Adorno has set out to exterminate it.[27] Heidegger's writings, which try to conceal their promiscuous relation to reactionary, "merely ontical" forces, are infected with the ideological thrust of a vocabulary that thrives on ambiguity. "Authenticity," a characteristic term in the jargon that Heidegger shared with many politicians, theologians and conservative ideologues, abstracts from social causes of discontent by giving contemporary feelings of meaninglessness an ahistorical formulation. Heidegger shirks responsibility for the claim inherent in the word "authenticity" to be presenting a positive doctrine of the good life when he insists that he is using the word as a value-free technical term, even while exploiting its fascination. That the alleged meaninglessness of life invalidates all principles of how to live, serves only to attract people to a certain way of life. Adorno analyzes this process whereby the concepts of the jargon manage to give the pretense of dealing radically with the crucial issues of life, society and philosophy, while they merely substitute the aura of connotation-laden words for the required content. Their false appearance has, according to Adorno, led to the surprising appeal of Heidegger's *Being and Time* and of the existentialism which it encouraged.

Reading Adorno, on the contrary, it is easy to be initially unimpressed. His style aims precisely at avoiding such thoughtless adherence to thoughts. Yet, what Adorno has to say has much of the urgency which in Heidegger's writings tends to be illusory. Adorno's critique of Heidegger, which cuts away the cancerous jargon to save the concerns that have become self-defeating, is of particular relevance to the attempt to learn from Heidegger and Marx together. *The Jargon of Authenticity*, oriented around Adorno's and Heidegger's comparative sensibilities to language, stands as a prolegomenon to the confrontation between the two mainstreams of twentieth-century continental thought.

Adorno's Early Criticisms

The forty-odd years of Adorno's stance towards phenomenology and Heidegger began in his student years, forming the basis for some of his first conversations with Horkheimer and Benjamin and culminating when he was twenty in a doctoral dissertation on Husserl. The critique of Husserlian phenomenology was later developed in more dialectical terms in *Zur Metakritik der Erkenntnistheorie*, which attacks the roots of many problems Adorno pointed to in Husserl's student, Heidegger. Adorno's first book, turning to another major influence on Heidegger, presents a rebuttal to existentialist ontology oriented on Kierkegaard, source of Heidegger's existentialism. Schroyer's forward to the translation of *Jargon* draws out this last connection.

Perhaps most interesting of Adorno's early writings is a series of three essays composed shortly after the publication of Heidegger's *Being and Time* but only recently made available in Adorno's posthumous collected works. The first, a programmatic inaugural address on *The Relevance of Philosophy* delivered in 1931 when Adorno began teaching, reflects upon the contemporary situation of philosophy by evaluating the failings of the various schools of the day. This lecture is striking both in terms of the importance it attributes to Heidegger and the thoroughness with which it sees through his pretenses. Adorno deals here with three instances of the necessary failure of Heidegger's accomp-

---

[27] Theodor W. Adorno, *The Jargon of Authenticity*, tr. K. Tarnowski & F. Will (Evanston: Northwestern University Press, 1973). Cf. Theodor W. Adorno, *Jargon der Eigentlichkeit* (Frankfurt: Suhrkamp, 1965).

lishments to live up to the promises of his rhetoric: *Being and Time*'s pathos of a radical new beginning is rejected by placing its problematic firmly within the context of the impasses reached by Simmel, Rickert, Husserl and Scheler; Heidegger's ethos of anti-idealistic concreteness is shown to be betrayed by his systematic method and presentation in *Being and Time*; the flaunted escape from subject-object metaphysics is understood by Adorno as a reduction to pure subjectivity.

Adorno's paper on *The Idea of Natural-History*, delivered a year later, views Heidegger's concept of "historicity" – one which instantly grates on Marxist nerves – as a false reconciliation of nature and history, of eternal structures and contingent facts. For the ontological theory of history can only achieve an adequate interpretation of Being if it foregoes such orientation toward structures of possibility in favor of a radical exegesis of the actuality of beings in terms of their determinations within concrete social history.

Finally, Adorno's *Theses on the Language of the Philosopher* criticizes Heidegger's linguistic novelties in terms of the historical conditions on philosophic prose. According to Adorno's theory, Heidegger's terminological innovations flee from history without escaping it. Heidegger exploits a highly situated jargon as though it had ahistorical validity and absolutizes historical concepts within a destiny of Being, which is unaffected by the social context. Consequently, despite his fondness for word plays and etymologies, his praise of the poets and his worship of language as the historical medium of being, Heidegger is accused by Adorno of lacking an aesthetic sensitivity to the social content of language, and this failing leaves him susceptible to the enticements of the jargon of authenticity and its unreflected provincialism. Anticipating the tack of *Jargon*, Adorno's essay on language concludes that "all deceptive ontology is to be exposed by critique of language."[28]

Dialectic of Enlightenment, written with Max Horkheimer during the war, exhibits significant parallels to Heidegger's writings, although it never refers to existentialism, ontology or their prime spokesman. The project of tracing the concept of reason (scientific enlightenment, *Vernunft, ratio, logos*) from the pre-Socratics to the technological age in terms of literary and philosophical texts is as central to Adorno's attempt to grasp the contemporary form of rationality, which had culminated in fascism, as to Heidegger's essays of the same period which share that goal. This comparison suggests that the conflict expressed in *Jargon* is not a matter of disparate world views hurling insults, but that despite his polemical tone Adorno agrees with Heidegger on the present concerns of philosophy as well as on certain methodological issues.

---

[28] Theodor W. Adorno, "Thesen über die Sprache des Philosophen," *Gesammelte Schriften I* (Frankfurt: Suhrkamp, 1973).

Yet there are crucial differences. The thesis which the *Dialectic of Enlightenment* substantiates, that the historic process of subject-formation has been accompanied by a de-subjectification through social forces and relations since time immemorial, is an implicit argument against ontology, whose concepts of man and Being cannot deal with the essential interpenetration in social history of that which these ontologized concepts leave abstract.

Adorno's Methodology of Critique

That Adorno relates the development of rationality, the relationship of myth to enlightenment, and various other concerns which he shares with Heidegger to Marx's analysis of capitalist relations of production while Heidegger maintains a strict primacy for the evolution of the ontological categories, indicates that Adorno was speaking for himself as well when he described Benjamin's attitude toward Heidegger. Noting Benjamin and Heidegger's shared rejection of idealist abstractions and formal generality, Adorno emphasized, however, that "the decisive differences between philosophers have always consisted in nuances; what is most bitterly irreconcilable is that which is similar but which thrives on *different centers*; and Benjamin's relation to today's accepted ideologies of the 'concrete' is no different. He (Benjamin) saw through them as the mere mask of conceptual thinking at its wit's end, just as he also rejected the existential-ontological concept of history as the mere distillate left after the substance of the historical dialectic had been boiled away.[29]

Adorno seeks to uncover the "center" on which Heidegger's analyses and their popularity thrive, for this center gives form and significance to the configuration of Heidegger's insights. The comprehension of the relation of this center to society – and not *directly* Heidegger's personal activity or class origins – provides the basis for a political judgment of Heidegger's philosophy. This approach is characteristic of Adorno's critical practice. According to his aesthetic theory, for instance, it is not the correspondence of individual contents of a work of art to specific social influences which accounts for the progressive or reactionary character of that work, but the way in which the work responds to prevailing social relations. Thus, in a letter to Walter Benjamin, Adorno writes, "I regard it as methodologically unfortunate to give conspicuous individual features from the realm of the

---

[29] Theodor W. Adorno, A portrait of Walter Benjamin," *Prisms* (London: Neville Spearman, 1967), p. 231.

superstructure a 'materialistic' turn by relating them immediately and perhaps even causally to corresponding features of the infrastructure. Materialistic determination of cultural traits is only possible if it is mediated through the *total social process.*"[30]

Adorno's philosophical interpretations proceed by the same maxims. Heidegger's work is treated neither simplistically nor deterministically; it is neither rejected out of hand as mere bourgeois ideology nor uncritically accepted as autonomous contemplation. It is comprehended, rather, as an arena from which the forces at work throughout society are scarcely excluded and in which any truth that manages to make an appearance will necessarily be conditioned by those forces – in one way or another.

Clearly, the penetration of social relations into Heidegger's system can only be revealed through a thorough grasp of the philosophical propositions, but these are not taken as ends in themselves: between the lines a social force-field must be reconstructed. In a tribute to his boyhood friend, Siegfried Kracauer, Adorno summarizes this approach to philosophical interpretation: "If I later, when reading the traditional philosophical texts, let myself be less impressed by their unity and systematic coherence, but rather concerned myself with the play of the forces which worked on one another under the surface of each closed doctrine and considered the codified philosophies as in each case force-fields, then it was certainly Kracauer who inspired me to it."[31] More than anything else, this oblique approach to philosophies – especially apparent in *Jargon*, which relates Heidegger to society in terms of the medium of a politically-loaded language-game – makes Adorno's critique of Heidegger difficult to grasp.

Adorno's Mature Criticisms

For years Adorno avoided the frontal attack on Heidegger anticipated in the early essays. The systematic intention of *Dialectic of Enlightenment*, probably to be attributed to Horkheimer, was uncharacteristic of Adorno. He spent his most productive years composing focused essays. Numerous references to Heidegger are sprinkled throughout these studies; the important discussions

---

[30] Theodor W. Adorno, letter to Benjamin dated 10 November 1938, *New Left Review*, October 1973, No. 81, p. 71.

[31] Theodor W. Adorno, "Der wunderliche Realist: Über Siegfried Kracauer," *Noten zur Literatur III* (Frankfurt: Suhrkamp, 1965).

of Kafka and Beckett, for instance, interpret their subject matter as poetic critiques of Heidegger, in explicit renunciation of the popular existentialist readings. When, near the end of his life, Adorno did present his conception of philosophy systematically, Heidegger was there front and center. *Negative Dialectics*, the only extensive mature work completed unless one counts the monograph on Alban Berg, devotes the first of its three parts to Adorno's "relation to ontology," a critique of Heidegger which provides the starting point for Adorno's own "anti-system." Perhaps the most significant contrast of Heidegger and Adorno would be one based on the latter's posthumously published *Aesthetische Theorie*. Such a study would, however, have few explicit connections to draw upon. Informed by the philosophical debates, it would have to note the shared rejection of subjectivistic aesthetics and evaluate the relation of art to society in the respective theories. Short of this, *Negative Dialectics* and its offshoot, *The Jargon of Authenticity*, will have to be accepted as the definitive statements of Adorno's critique of Heidegger.

According to the introduction to *Negative Dialectics*,[32] the task of philosophy in our times is the transformation of subjectivistic thinking by means of the subjective strength of the critical individual. The subsequent priority of substance over the knowing subject implies a primary concern with the concrete, which has been distorted under the demands of a coercive social totality. Although method would then be determined by the subject matter, analysis could not proceed without concepts. This linguistic requirement presupposes a critique of the philosophical tradition, that is, of German idealism and of the inept criticism of idealism by positivism, phenomenology and existentialism. While these goals may capture much of Heidegger's stated intentions, according to Adorno's account, Heidegger, like Husserl before him, has failed to deal adequately with the complexities involved in grasping the concrete.

In *Negative Dialectics* Adorno suggests how the concrete is missed by Heidegger's simplistic scheme, which underlies and supports an elaborate obscurantism. The three poles of Heidegger's system – beings, human existence and Being – interpenetrate each other only formally, without taking into account their configuration, which defines their content. The concrete social history in which these poles, as dialectical, intertwine and develop according to Hegel and, in effect, Marx, disappears in Heidegger's presentation. Thereby their present forms are not clearly situated in history; as essential and eternal, they are, thinks Adorno, glorified and affirmed. The oft-bemoaned quietism of Heidegger's later writings is thus revealed by Adorno to be non-accidental: it is a consequence of the very approach of the ontological project, one which excludes social content from the start.

---

[32] Theodor W. Adorno, *Negative Dialectics* (New York: Seabury, 1973).

This criticism is particularly interesting because Adorno has also been accused of *praxis* paralysis and because Heidegger can respond as Adorno has that his emphasis on contemplation is a reaction against a preponderance of thoughtless pragmatic activity in present society. The difference between the two philosophies is that receptivity becomes a dead-end in Heidegger's system, rather than a corrective moment, which negates only the distortions and limitations of social practice. The philosophical source of the difference is that Heidegger's approach reacts too simplistically to the dilemmas of post-Hegelian philosophy, attempting to skirt the problem of a non-idealistic mediation of subject and object, of thought and society, of theory and practice. Where Adorno radicalizes Hegel's dialectic, redefining it in terms of the non-identity of word and object and articulating the mediations involved more thoroughly than even Hegel, Heidegger falls behind Hegel, hypostatizing language along with Being outside the influence of that reality which they characterize.

This theoretical point has practical consequences for Heidegger's philosophy insofar as he fails to reflect on the relation of society to his language. Heidegger's failure to deal adequately with the present social context of philosophy is perhaps Adorno's strongest indictment of him: his ontology is an unfortunate response to social conditions in which people feel powerless. In the guise of a critique of subjectivistic will, it fetishizes the illusion of powerlessness and thereby serves those in power. Following a restorative thrust, Heidegger's formulation of a real felt need merely assumes a solution and thus serves to perpetuate the underlying problems according to Adorno's analysis. Strengthening conservative ideology, Heidegger's approach avoids those issues that point to the realm of society, an arena in which people could possibly exert some joint control.

The Jargon of Authenticity is more focused. Unlike *Negative Dialectics*, which addresses itself to the central *topoi* of Heidegger's thought as a whole, *Jargon* seems to limit itself to an area of questionable importance, although it brings an impressive array of considerations to bear. Dealing with Heidegger's pivotal "question of Being" only peripherally, it is preoccupied by the accompanying doctrine of man. Further, it zeroes in on terms and themes which Heidegger himself dropped after *Being and Time*. Thus, of the four sections of Adorno's essay (beginning on pages 3, 49, 92 and 130), the first reflects on the jargon in the hands of Heidegger's predecessors, colleagues and followers, barely mentioning Heidegger himself. The next section fits Heidegger into this picture, but notes that Heidegger protects himself against the imputation of the jargon's worst offences even while exploiting its appeal. Another part is devoted to the concept of authenticity, which Heidegger never again used so freely after the reaction to his first book. In the final pages, the choice of the analysis of death as an illustration of Heidegger's procedure involves Adorno in the non-intuitive argument that people might overcome death in a future social arrangement. Even if this is possible – and in *Jargon* it remains an empty possibility – Heidegger has still articulated the importance of finitude as an essential feature of the human condition as we know it. Concentrating as he does on the social consequences of Heidegger's concepts of authenticity and death, Adorno seems to miss the role these play in Heidegger's ontology. For authentic Being-towards-death is less a moral stance in Heidegger's system than a condition of the possibility of valid ontological reflection.

Jargon thus seems open to the very criticism it levels against *Being and Time*, namely that the pragmatic impact on the reader is not substantiated by the propositional evidence. Just as the popularity of Heidegger's work was attributed by Adorno largely to moral connotations explicitly excluded from the epistemological discourse, so it seems that Adorno's own essay gives the impression of utterly destroying Heidegger's philosophy when it merely picks at incidental themes without understanding their import.

Viewed from the perspective of *Negative Dialectics*, however, Heidegger's analysis of human existence is symptomatic of his later investigations of tool, artwork, thing and word, even of Being itself. Although the structures of man, thing and Being include, on Heidegger's account, relations to each other, the concrete social history in terms of which they affect each other through these otherwise abstract relations is left out of consideration. This fault can be demonstrated just as meaningfully in terms of Heidegger's early *Daseinsanalytik* as with the later *Seinsfrage*, and the political implications that follow from either are more clearly drawn out of the former. In short, *Jargon's* oblique social attack on the linguistic aspect of a supposedly moralistic part of Heidegger's early thought succeeds in making thoroughly problematic many central characteristics of Heidegger's approach and system in general.

Significantly, Adorno's social critique of Heidegger is not simply divorced from a philosophical one. Rather, it underscores the philosophical failure of Heidegger's thought: its lack of concern for the very social dimension in which it becomes self-defeating. This particular failure necessitates the confrontation between Heidegger's and Marxist critical theory of society. By determining the social limitations of Heidegger's thought, Adorno does not discard Heidegger, but attunes the strivings of Heidegger's philosophical concepts to their social content, measuring the distance between their claims and their achievements. Only thereby can Marxism interpret Heidegger's insights within the context of Marxism's own method and fruitfully comprehend both the progressive and the reactionary force of Heidegger's socially-situated path of thought.

PART II. KARL MARX: IDEOLOGY CRITIQUE AS INTERPRETATION AND TRANSFORMATION OF THE WORLD

Karl Marx developing revolutionary theory.

Chapter IV. Anticipations: The Early Works

Marxism is not just one more show in a repertoire of philosophical fantasies. When the curtain rises, it is not to reveal a self-contained, harmonic production, but rather to present the analysis of the drama which unfolds itself outside. No passive entertainment, the show charges admission: active commitment to a better world. The line, "something is foul in the state of Denmark," is not discovered mid-way, but presupposed by the whole. Marxism is the philosophy of revolution, stating the preconditions for change, but not deducing its desirability as though it required proof. Construed as a system, Marx's thought presupposes only the self-contradictory social reality which it articulates.

Insofar as one distinguishes Marx's thought from the social system he criticizes, the presupposition of the latter by the former can be viewed in two ways. Marx's methodology is a response to the contradictory character of capitalist society, while his interpretation of that society as fundamentally contradictory is related to his critical approach. This paradoxical or, perhaps better, circular structure to Marx's project defines its uniqueness. Rejecting the relativism of manifold worldviews – which are at any rate unconsciously conditioned by social relations – Marx develops his outlook through research into social objectivity. At the same time, he avoids pre-critical metaphysics by realizing that the comprehension of capitalist society presupposes a theoretical framework. The priority of social existence over thought, due at least to the fact that any thinker finds himself always already within a social structure which he did not create, rescues from the charge of arbitrariness even Marx's most basic critical intention, itself a consequence of the inadequacies of the society in which he found himself. The partial identity, through mediation, of subject and object, of social critic and capitalist society, manifests itself in the tendential identity of the theoretician's social critique of theory and capitalist society's theoretical critique of its own contradictory character.

The anti-philosophical remarks of the early Marx outline only a fraction of the extensive system of thought they implicate; they represent the mere peak of an iceberg kept submerged of late in the murky waters of cold-war rhetoric. Broken off of the whole, an isolated sampling of Marx's thought must dissolve like an ice cube in an eclectic cocktail of ideas, watering down the potency while pretending to revolutionize. The point is to comprehend the unity of Marx's thought in its relation to social reality and practice. Dependent upon its social context for the content of its presuppositions, Marxism nevertheless carries out its analysis autonomously, using the historically given categories and contradictions to transcend their own apparent limitations. The results stand as a condemnation of existing social relations, but not as a merely moral disgust. Rather, truth and falsity have been separated out of ideology, concepts have been re-forged through criticism, the cores of problems have been exposed and potentials for rectification have been revealed.

The guiding theme of the following interpretation of Marx is that he places the concept of commodity production at the center of his theory of bourgeois (capitalist) society, with profound consequences for the philosophical trappings of this theory. Such an understanding of Marx is in conscious opposition to several prevalent tendencies in the secondary literature. Too often Marx's youthful references to alienation in the *Manuscripts* are sighted as moralistic, psychologistic or idealistic – at any rate, crudely separated from the mature economic analyses. The notion of *praxis* in the *Theses* is hypostatized into the basis of an ontology of *praxis*, as though the concept was not designed in this context precisely to attack the ahistorical conceptualizations of Feuerbach. In contrast to those philosophically speculative interpretations, other discussions take Marx's political economy to be merely an empirical science whose validity stands and falls solely in the comparison of its isolated parts to competing hypotheses in that science: from Adam Smith to Paul Samuelson. The historical footnotes to *Capital* – most extensively elaborated in the section on "Forms Which Precede Capitalist Production" in the *Grundrisse* – are taken by many to be identical either to Hegelian universal history, thought to reduce history to an empty tripartite schema, or to naive historiology, the attempt to restate "that which was." It is necessary to oppose these procedures of divide and conquer which tear the unity of Marx's thought asunder along chronological lines by distinguishing Marx the philosopher, sociologist, economist, revolutionary, humanist and private citizen, thereby, intentionally or not, reducing each misunderstood segment to meaninglessness under the pretext of saving it from a questionable whole.

The opposition to the fragmentation of Marx's work receives its justification in the essential unity of Marx's underlying purposes throughout his writings. in focusing on commodity production, the interpretation emphasizes that cornerstone of Marx's analyses which is implicitly intended with the terms "alienated labor" and "*praxis*," explicitly with those of "production," "free labor" and "commodity fetishism." Accordingly, the two most popular early writings, *Alienated Labor* (August 1844) and *Theses on Feuerbach* (March 1845), will be interpreted by, in effect, introducing the concept of commodity production into the discussions which are couched in more ambiguous terms: alienated labor and social *praxis*. Through this paraphrasing, Marx will be shown to be arguing similar points in his early, philosophical works as in his mature, economic writings, subsequently to be considered. The "Introduction" (August/September 1857) to the *Grundrisse* presents the most explicit argument for using the concept of commodity production as the basis for an understanding of the capitalist system. The historical account in the *Grundrisse*, the chapter on "Forms Which Precede Capitalist Production" (December/January 1858), is not a history in the normal sense, but a retrospective account of the development of the material preconditions of commodity production.

Finally, it will merely be necessary to round out the picture of the unity of Marx's work with a glance at *Capital* (1867), which recapitulates the by now familiar themes in the terms of a fully elaborated system. Here the commodity is analyzed into its two aspects: use value and exchange value; the contradiction between these is shown to be at the root of fetishism; and the theory of surplus value, which also derives from this contradiction in commodity production, is elaborated into a theory of industrial society. Of particular interest are the relation Marx establishes between his procedure of abstraction and developments in society; the ontological implications he draws from the commodity character of the products of capitalist production; and the use of the results of these analyses for his demystification of fetishism.

The Primacy of Commodity Production for Interpretation

The joy of liberal revisionism and the embarrassment of reactionary dogmatism find their source in Marx's early manuscripts, texts which therefore provide a natural starting point for a contemporary evaluation of Marx's foundations for social theory. Marx's popular discussion of alienated labor, when carefully viewed as an analysis of commodity production, is a first major document of his "turn" from philosophy and law to political economy as the object of ideology critique. Although not yet the explicit center of attention, the concept of the commodity is already present, to be developed in the succeeding thirty years in terms of the contradiction between use value and exchange value, the labor form of value and the appropriation of surplus value. Moreover, Marx's method of starting from the given reality and the prevalent ideologies to develop his own conceptualizations is clearly at work, despite the misleading form of logical (dialectical) derivations from the concept. A further cause for confusion is Marx's penchant for adapting traditionally metaphysical terms to an anti-metaphysical project. (This procedure is not without its justifications, but the danger of misinterpretation is enormous, as is clear from the way it confused Heidegger, who had himself used the technique against metaphysics.) With these dangers in mind, we turn to Marx's most controversial ten page text.

The first of the *Economic-Philosophic Manuscripts* (1844) is divided under the traditional headings of political economy: "Wages of Labor," "Profit of Capital" and "Rent of Land," with a concluding section referred to as "Alienated Labor." This last section begins by summarizing the procedure and results of what preceded:

> We have proceeded from the presuppositions of political economy. We have accepted its language and its laws. We presupposed private property, the separation of labor, capital and land, hence of wages, profit of capital and rent, likewise the division of labor, competition, the concept of exchange value, etc. From political economy itself, in its own words, we have shown that the worker sinks to the level of a commodity, the most miserable commodity; that the misery of the worker is inversely proportional to the power and volume of his production; that the necessary result of competition is the accumulation of capital in a few hands and thus the revival of monopoly in a more frightful form; and finally that the distinction between capitalist and landowner, between agricultural laborer and industrial worker, disappears and the whole society must divide into the two classes of *proprietors* and property-less *workers*.[33]

Marx's procedure is a dialectical form of "immanent critique." In criticizing the theories of bourgeois political economy, Marx does not attack from an alternative position based on its own set of presuppositions, but starts out from the most highly developed form of that theory itself. From within the questionable position, Marx subjects its concepts, suppositions and analyses to a form of critical self-reflection, calling them into question by relating them to each other, seeking their origins, and further developing them into a *reductio ad absurdum*. Thus, "just exchange" leads by a development of its very logic to its opposite: an inverse relationship between the worker's wealth and the value of his products; competition likewise results in monopoly; and, finally, the separation of capital and land-ownership develops into a unified class of proprietors.

Where has political economy gone wrong, then, according to Marx? It starts from the fact of private property, but it establishes this fact in abstract laws

---

[33] Karl Marx, "Alienated labor," *Economic and Philosophic Manuscripts (1844)* in *Writings of the Young Marx on Philosophy and Society* (Garden City: Doubleday, 1967), p. 287. Cf. Karl Marx, Die entfremdete Arbeit," *Okonomisch-Philosophische Manuskripten (1844)* in Karl Marx, *Texte zu Methode und Praxis* (Reinbek: Rowohlt, 1966, Bd. II, S. 50f.

rather than analyzing it in terms of (1) related concepts, (2) its historical origin or (3) its implications:

> Political economy proceeds from the fact of private property. It does not explain private property. It grasps (*fassen*) the actual, *material* process of private property in abstract and general formulas, which it then takes as *laws*. It does not *comprehend* (*begreifen*) these laws, that is, does not prove them as proceeding from the nature of private property.[34]

In terms of the distinction drawn in Chapter I above between explanation and interpretation (Marx uses the terms grasp and comprehend), traditional political economy has limited itself to *explaining* market phenomena in terms of a set of laws and concepts, without bothering to *interpret* these laws and concepts within an encompassing theory of society so that the laws and concepts could be comprehended as elements of a meaningful whole. Of course, the political economists did have a world view in terms of which they interpreted their economic theory, but this bourgeois ideology (in the strictest sense of the term) was merely superimposed on the science: it did not grow out of the relationships within the science. Further, both the ideology and the science lacked reflection upon their historical limitations. Marx's originality lay in the unifying of explanatory science and interpretive framework. Incorporating historical reflection at the heart of this unity, Marx's thought becomes "theory," lacking the arbitrariness of world view and the provinciality of ideology.

Marx's dialectical "comprehension" of the phenomenon of private property can be conceived as a three-pronged attack: (1) He sets out to "grasp the essential *connection* among private property, greed, division of labor, capital, ..." (2) He wants to show how the concepts and relationships of capitalism are "necessary, inevitable, natural consequences" of an historical *development* whose previous stage of feudalism incorporated monopoly, the guild and feudal property. (3) As a result of these analyses, he hopes to *dispel the ideology* of private property, just exchange, individualism. Marx's strategy – in the *Manuscripts*, as in later writings – is *to show that commodity production is the essential determinant*, specifying the historical content of property, exchange and the individual in bourgeois society.

Anticipating this priority of the productive realm, Marx starts from a fact about production, rather than about property, later deriving private property as a consequence of capitalist relations of production. Rejecting the traditional starting point of philosophy with some unconditioned concept or

---

[34] *Ibid.*, p. 287, S. 51.

situation – Hegel's "Being" or political economy's "state of nature" – Marx proceeds from a "political-economic, *present* fact." Not only is this an economic fact visible in the contemporary society – as was the political economist's fact of private property – but it is a theoretical "fact" at this stage of analysis in Marx's *Manuscripts,* the result of his preceding immanent critique of the theories of the political economists. The fact, a moral indictment of capitalism, a contradiction in its ideology and an indication of its severe limitations, is formulated as follows:

> The worker becomes poorer the more wealth he produces, the more his production increases in power and extent. The worker becomes a cheaper commodity the more commodities he produces. . . . labor not only produces commodities. It produces itself and the worker as a *commodity.*[35]

That such a "fact" provides the basis of Marx's theory in his manuscript is revealing of its anti-epistemological attitude. The belief that no theory of knowledge can independently precede the theory of society is premised upon a conclusion Marx later elaborated in *The German Ideology*: the sensuous world is the product of industry and societal conditions; it is an historical result. Rejecting Kant's emphasis on ahistorical and non-social categories of pure reason, Marx saw the condition of the possibility of knowledge, whether scientific or everyday, in the present totality of social practice. More in agreement with Hegel that knowledge – as the mediation of subject and object with the help of concepts – is itself mediated by the object rather than merely structured by an unmediated consciousness, Marx nevertheless rejected Hegel's concern with the concept as origin and result of the historical process. The concept is relegated by Marx to the position of an intermediate moment in the comprehension of socio-historical reality. Marx's theory presupposes an objective world already mediated by historical human activity. His goal is to comprehend this world by conceptually articulating the mediations that define it. For these reasons, Marx begins with a fact of present reality. That this fact is known *a posteriori*, does not leave its choice arbitrary. Anticipating the analysis to follow, it is chosen centrally so as to provide essential categories that can be elaborated in the final theory. Consequently, as subsequent research reveals new insights, the starting point of the presentation must be modified.

The fact from which Marx starts in the manuscripts is historically specific, not valid for all times, but this does not mean it was chronologically limited to Marx's time and perhaps explainable by accidental circumstances of

---

[35] *Ibid.*, p. 289, S. 52.

politics or inflation. Rather, the fact is stated with complete generality. Its specificity first becomes apparent when one notes that workers did not always produce large amounts of "wealth" – traditional peasants produced mainly food for themselves and their dependents. The worker has not primarily produced "commodities" in all epochs, previously he produced mostly "use values" for immediate consumption, rather than "exchange values" for sale. Marx's fact is thus just as timely as *commodity production* itself as the dominant form of production. When Marx refers to production in his essay, he is thinking of commodity production; labor is wage labor which produces commodities in exchange for money; products are commodities made solely for the purpose of being sold; private property is the ownership of the material preconditions of commodity production; and wealth is abstract value, the monetary worth of commodities. These concepts of political economy are thus components of a theory of *capitalist* society. Their content is defined by the relationships that inhere in this system as distinguished from other systems, such as the feudal society out of which it grew. This already suggests the importance of Marx's three-fold procedure: (1) To grasp the systematic connections among the concepts is not only part of what it means to analyze the concepts themselves, but also a necessary step in relating the concepts to social reality. (2) Tracing the logic of development of the concepts as aspects of the real social system dispels the myth of eternality, suggesting future contradictory developments and a perspective for revolutionary change. (3) The first two steps involve a rejection of liberal ideology, which ignores and distorts the conceptual interrelationships and enthrones the concepts in an aura of eternality and necessity.

By carefully distinguishing capitalism from previous social systems and drawing the consequences of this distinction for theories of society, Marx transcended Hegel and Adam Smith simultaneously and criticized both economic theory and practice in one blow. For it then became clear that production of commodities is "alienated" labor not merely in the Hegelian sense of a subject externalizing himself in the objective world – applicable to all forms of society and scarcely to be transcended outside the realm of mind – but in the capitalistic sense that the product is someone else's property. With this all-important twist, the dialectic of consciousness with nature, its recognition of the other and its coming to self-awareness are essentially altered. It can then be seen that Hegel is just as guilty as the economists, from whom he gleaned much of his material, of treating certain bourgeois categories, such as private property, without concern for their social specificity and therefore without awareness of their actual content. The apologetic nature of political economy, precisely in its utopian ambitions, can be attributed to its absolutizing of capitalist categories, judging all past and possible social forms by the standards inherent in capitalism alone, and necessarily concluding that capitalism is the most "rational" (in this context itself an enlightenment value with specifically capitalistic content). Only by carefully distinguishing that which is unique to capitalist reality and its conceptualizations, can Marx reveal how capitalism entails both new emancipatory potentials and increased repressions.

The Alienated World

Marx's often puzzling remarks and rambling style in the *Manuscripts* can be greatly clarified simply by bearing in mind that alienated labor always refers to commodity production. Marx literally underlines this when he writes: "*Political economy conceals the alienation in the nature of labor by ignoring the direct relationship between the laborer* (labor) *and production*."[36] Marx traces the effect upon the laborer of the fact that his labor is part of the process of commodity production; the whole analysis of alienation follows from this. Centering on the laborer, Marx successively shows that, as a result of alienation from his product as something that will neither belong to him nor be consumed by him, the laborer is alienated from (1) nature, (2) himself, (3) humanity, and (4) other people.[37] These manifold forms of alienation are results of the "present fact of political economy" from which Marx starts out. The contradictory

---

[36] *Ibid.*, p. 291, S. 54.

[37] This is Marx's list in his manuscript. Since 1844, commodity relations – originating in the relation of the laborer to his product as a commodity – have spread throughout all sectors of society. Marxist cultural criticism (Lukacs, Benjamin, Adorno, e.g.) has, for instance, more recently analyzed the consequent effect on art. The recent emphasis upon consumption, which some would argue makes Marxism obsolete, can itself be traced to the relations of production, as Enzensberger suggests in his reply to the argument that ecological considerations call for moderation by consumers (Hans-Magnus Enzensberger, "Critique of political ecology," New Left Review, No. 84, p. 30; Cf. the quotation from Marx in footnote 3 above):

> To ask the individual wage-earner to differentiate between his 'real' and his 'artificial' needs is to mistake his real situation. Both are so closely connected that they constitute a relationship which is subjectively and objectively invisible. Hunger for commodities, in all its blindness, is a product of the production of commodities, which could only be suppressed by force.

relationships in which the wage laborer becomes entangled stem from the central contradiction: commodity production. Production in general is the appropriation (*Aneignung*, making one's own) of nature, but commodity production specifically is making something someone else's – the capitalist's and then the purchaser's – thus alienation (*Entfremdung*, making strange). These contradictions are not mere figments of logic, but historical results.

(1) Viewed historically, the *introduction* of commodity production into an economy based on self-sustaining economic units (extended family or estate) with marginal trade started a process which led to the dissolution of those units, the establishment of a commodity-based economy and the progressive impoverishment of the workers relative to the social standard of living. The more the worker turned from the direct satisfaction of his specific needs to commodity production, the less he owned in the way of materials, equipment and even food. Increasingly, the worker had to turn to someone else who could supply these preconditions of labor and life. Eventually, the laborer was left with nothing but his physical labor power, and he was reliant upon selling that to keep alive. Separated from the land and hereditary estate, the worker became *alienated from nature* as source of the preconditions of his occupational and physical existence because these preconditions now belonged to someone else. This is one sense in which the worker becomes poorer the more commodities he produces: he becomes alienated from nature. This alienation is a result of commodity production itself as a mode of production and as a determinant of historical development.

(2) Another such result is the determination of the laborer himself as a commodity. No longer producing in accordance with his particular needs and desires, his activity, which he is forced to sell on the market, is usually controlled by someone else and is always aimed at satisfying needs of some unknown, hence abstract, other. Divorced from the source of the meaning of his own existence, his own activity, the laborer is *alienated from himself*. An illegitimate metaphysical distinction has been drawn – by society, not by thought – between the individual and his activity, where the latter belongs to some other individual. A pervasive schizophrenia has manifested itself in society, leaving a whole class of people with not two, but only half a personality.

(3) The fragmentation of the worker – not only into psychological ego and objective activity, but into cripplingly specialized fragments of activity – is particularly disturbing because it comes as the result of the process of progressive universality. Capitalism involved the dissolution of the traditional limitations upon human development and brought with it the technology to create universally, according to "intrinsic standards." The freedom potentially available to the individual as a result of his belonging to the "present and

living species" is, however, repressed by the form of commodity production, which reduces this "free spontaneous activity" to a mere means for the laborer's physical existence. Unable to benefit from humanity's great advances, the wage laborer is *alienated from* the very notion of *mankind*, an ideal unattainable, foreign to its members.

(4) In the end, the commodity-based economy reduces human communities to the "war of all against all" which political economy and bourgeois philosophy had projected back to society's origins. Alienated from his product, nature, himself and mankind, the laborer is necessarily *alienated from others*, whether they be the Capitalists who appropriate the product immediately, the purchasers who finally consume it or fellow workers who compete for jobs. The other, who once may have helped out for the common good or shared a common world, has in capitalist society become part of a system which tears the laborer's life and world asunder. The abstract nature of commodity relations, which reduce everything to monetary terms, pits the laborer concretely against society as the Fichtean Other.

Marx's classic presentation of alienated labor in its various aspects has not yet turned to the social context in which commodity production unfolds its consequences. In analyzing alienation, Marx has analyzed the structural relations between the laborer and his product in the process of commodity production and he has done this in terms of the effects upon the laborer. Next he must turn to the historic process in which commodity production establishes itself. Marx has only one further point to make, namely that the *existence of the capitalist* is a necessary corollary to alienated labor. After all, Marx reasons, if the product of labor is alien to its producer, it must belong to another person. Further, the laborer will necessarily relate to this other person in terms of the process of production as an "alien, hostile, powerful man independent of him," as the lord of his product.

The process of production thus creates, along with alienation in its various aspects, the relation in which the producer stands to the lord of labor, the owner of the materials and means of production, the designator of the details of production, the payer of wages for support. the owner of the product. "The relations of alienated labor, of commodity production, thus produce the relation of the capitalist to labor. *Private property* is thus product, result and necessary consequence of *externalized labor*, of the external relation of the laborer to nature and to himself."[38] Private property is another expression of capital, as opposed to the wages of labor. The point is that the accumulation of wealth (materials, equipment and money) in the hands of a few is a necessary consequence of commodity production by the masses. Political

---

[38] Cf. *ibid.*, p. 297, S. 60.

economy, by contrast, starts from the concept of private property as though it were an ageless notion and from an imaginary primordial situation in which most people had no property and therefore had to exchange their labor time for money to meet their needs. Wage labor was thus said to result from private property, if, indeed, a relation was ever drawn. This prevalent view justifies alienated labor, then, in terms of the ideology of just exchange. Marx's view, on the other hand, shows commodity production to be not only the basis of an exploitative system, but the original cause of the inequitable distribution of wealth as well – far from a rational solution to a "natural" inequality.

In true dialectical fashion, Marx not only builds his new view on the rubble of the ideological edifice he destroyed, but indicates as well how the mistaken appearance arose: "Only at the final culmination of the development of private property does this, its secret, reappear – namely, that on the one hand it is the *product* of externalized labor and that secondly it is the means through which labor externalizes itself, the *realization of this externalization*."[39]

We can reconstruct the history of private property as follows: Early, undeveloped forms of private property such as merchants capital provided a basis for commodity production to begin through organization of traditional home crafts into manufacturing systems. This heralded the social division of labor into property-less laborer and propertied non-laborers. Although presupposing the existence of private property in some elementary form, commodity production itself reproduced and developed its preconditions; it was a motor of the capitalist accumulation which bourgeois political economy projected back to an ahistorical state of nature in which the clever and physically powerful greedily took from the weaker people. Because commodity production historically presupposes private property, property was assumed to be systematically prior. In its developed form, however – as industrial capital – it is a result. The secret of private property is really a secret of commodities and their fetishism, for property is the fetishization of interpersonal relations. This is a problem whose explication must await the discussion of *Capital.* Here it is merely important to note Marx's characteristic insight that what is historically prior may at the end of its development retrospectively be seen to be logically derivative in terms of a contemporary conceptual framework of interpretation.

Marx draws two conclusions from his consideration of private property and the externalized labor of commodity production. First, he claims that one can develop all the categories of political economy with the aid of the concept of alienated labor and that each resultant category (like private property) will be

---

[39] *Ibid.*, p. 298, S. 60.

but a particular aspect thereof. Alienation, in the sense of the character of the relations of commodity production, is thus the analytic "essence" of the capitalist system. Marx's second conclusion is that the analysis of the historical roots of our society should be structured in terms of the relationship of externalized labor to the history of mankind, rather than as the traditional search for the origins of private property.

These two conclusions are central to Marx's later writings. The "Introduction" to the *Grundrisse* argues for the importance of the role played by the essential analytic concept and the opening chapter of *Capital* presents the concept of the commodity as such an essence in terms of which one can develop all the other categories of a theory of capitalist society. On the other hand, the section of the *Grundrisse* on the "Forms Which Precede Capitalist Production" traces the development up to and during the bourgeois era in terms of the relationship of the forms of labor to the progress of mankind. Finally, drawing on all this, *Capital* presents an historically specific analysis of capitalist production in terms of the form of value which corresponds to alienated labor.

If Marx's argument in the manuscript on alienated labor appears shaky or his vocabulary misleading, retaining concepts like "alienation" and "externalization" from Feuerbach and Hegel, whom he criticized, rather than explicitly referring to commodity production throughout, perhaps this explains his not publishing the manuscripts. Perhaps this is a reason for going on to the later formulations, rather than for rejecting Marx as idealistic, impressionistic, existentialistic. After a look at his *Theses on Feuerbach*, *de rigueur* for a consideration of his key concepts, we will pick up Marx's argument about commodity production in those sections of the *Grundrisse* and *Capital* referred to. They should provide adequate additional experience with his approach.

Ideology Critique and the Transformation of the World

Marx's *Theses on Feuerbach* deserve to be quoted in full and commented on individually:

> (1) The chief defect of all previous materialism (including Feuerbach's) is that the object, reality, sensuousness is conceived only in the form of the *object or image*, but not as *sensuous human activity, practice*, not subjectively. Hence in

opposition to materialism the *active* side was developed
abstractly by idealism, which naturally does not know actual,
sensuous activity as such. Feuerbach wants sensuous objects
actually different from thought objects: but he does not
comprehend human activity itself as *objective* activity. Hence
in *The Essence of Christianity* he regards only the theoretical
attitude as the truly human attitude, while practice is
understood only in its dirtily Jewish form of appearance.
Consequently he does not comprehend the significance of
"revolutionary," of "practical critical" activity.[40]

Production as Marx's analytic category is a synthesis of the constitutive
function and of sensuous perception, of Hegelian idealism and Feuerbachian
materialism. Social practice, defined by the prevailing form of production and
the corresponding social relations, is thus the activity by which mankind
constitutes the objective world in which *it* is actively situated, not merely the
adoption of a theoretical or passively receptive stance toward an external
world. Once this is recognized, it becomes clear that critical activity involves
actually changing the mode of production, a far cry from stopping with the
initial step of changing a few people's minds as did Hegel and Feuerbach.

> (2) The question whether human thinking can reach
> objective truth is not a question of theory, but a *practical*
> question. In practice man must prove the truth, that is,
> actuality and power, this-sidedness of his thinking. The
> dispute about the actuality or non-actuality of thinking,
> thinking isolated from practice, is a purely *scholastic* question.

Thought cannot meaningfully be divorced from the reality of social practice:
not only because that provides its object, but because the thinker is
historically situated. Thus, even in the case of a utopian vision, its truth lies in
the possibilities of its achievement, not merely in the abstract validity of its
arguments and values. (Cf. thesis #11 on the unity of theory and practice.)

> (3) The materialistic doctrine concerning the change of
> circumstances and education forgets that circumstances are
> changed by men and that the educator must himself be
> educated. Hence this doctrine must divide society into two
> parts, one of which towers above. The coincidence of the
> change of circumstances and of human activity or self-

---

[40] The Theses on Feuerbach are quoted in full from the *Writings* cited in
footnote 1 above and are compared with the German from the *Texte*, p.
400ff, S. 190ff.

change can be comprehended and rationally understood only as *revolutionary practice.*

The glimmers of hope twinkling in the future and the most progressive thoughts of the age have a common source: possibilities inherent in the past and present. There is no standpoint of objective knowledge on the other side of now and anyone who wants to help consciously create a better future – for the future will be created by people whether they consciously and democratically direct their efforts or let other people manipulate them – must educate himself about the society and potentials for change by situating his thought in the context of the revolutionary task of his society. Such self-education can take place only in unity with an activity which transforms the social circumstances in which all thought is founded. Transformative practice, if it is to be self-conscious, necessarily transforms the theory of society along with the mode of production.

> (4) Feuerbach starts out from the fact of religious self-alienation, the duplication of the world into a religious and secular world. His work consists in resolving the religious world into its secular basis. But the fact that the secular basis becomes separate from itself and establishes an independent realm in the clouds can only be explained by the cleavage and self-contradictoriness of the secular basis. Thus the latter must itself be both understood in its contradiction and revolutionized in practice. For instance, after the earthly family is found to be the secret of the holy family, the former itself must then be theoretically and practically nullified.

The key to uncovering forgotten potentials is the ideology critique of repressed embarrassments. Non-radical criticism is satisfied to stop after exposing the embarrassment in the *first* of the five steps which constitute ideology critique: (a) Demystify; reveal the ideology to be a false consciousness. (b) State the social causes of the rationalization; uncover its social necessity. (c) Construct a theory of society from the hints given by the ideology's moment of truth and by the specific necessity of its false aspect. (d) Analyze the possibilities of changing the society in terms of eliminating the social contradictions, which necessitated the ideology and redirecting the energies which it sublimated. (e) Follow the inherent demands of the ideology critique to change society at its roots; transform the essence of man, human practice.

Religion, as an opiate, not only acts as a narcotic in blurring the perception of social domination (by despot, aristocracy, king or capitalist) by substituting the illusory image of God, but it stands as a symptom in which critical thought can discern an oppression of consciousness necessitated by economic enslavement. No longer worshipped in their sublimated image, the chains which are the proletariat's only possession become an abomination to their captives. The struggle for emancipation is the natural consequence of an ideology critique founded in social objectivity and encouraged to run its course.

> (5) Feuerbach, not satisfied with *abstract thinking*, wants *perception* but he does not comprehend sensuousness as *practical* human-sensuous activity.

The rejection of the idealist's transcendental ego cannot rest content with placing the mind in a body conceived of in ahistorical-biological terms. As essentially practical activity, sensuous perception takes place in an historically specific social context.

> (6) Feuerbach resolves the religious essence into the *human* essence. But the essence of man is no abstraction inhering in each single individual. In its actuality it is the ensemble of social relationships. Feuerbach, who does not go into the criticism of this actual essence, is hence compelled (1) to abstract from the historical process and to establish religious feeling as something self-contained and to presuppose an abstract – *isolated* – human individual; (2) to view the essence of man merely as "species," as the inner, dumb generality which unites the many individuals *naturally*.

The essence of man is in each instance related to the present and living species, defined by its social practice. The essence is not to be sought in the biological specimen, but in the ensemble of social relations. Social theory's question, What is man?, today finds its answer in the capitalist relations of production, not in supposedly eternal abstractions.

> (7) Feuerbach does not see, consequently, that "religious feeling" is itself a social product and that the abstract individual he analyzes belongs to a particular form of society.

The individual – underlying substrate for ahistorical definitions of man – is itself a product of historical development and ideological sophistication.

> (8) All social life is essentially *practical*. All mysteries that lead theory to mysticism find their rational solution in human practice and the comprehension of this practice.

The comprehension of bourgeois social practice as the origin of fetishism is the key to untangling the confusions and illusions surrounding social existence.

> (9) The highest point attained by perceptual materialism, that is, materialism that does not comprehend sensuousness as practical activity, is the view of separate individuals and civil society.

The highest stage reached by philosophy before Marx, or by those subsequent philosophies which did not learn from him, was still restricted by the limitations of bourgeois society and the controlling interests of the bourgeoisie, for it did not reflect upon the effects of its social preconditions.

> (10) The standpoint of the old materialism is civil society; the standpoint of the new is human society or socialized humanity.

In transcending the limitations of bourgeois philosophy by reflecting upon the limitations and contradictions of bourgeois society, Marx strives for a society in which conflicting class interests have been resolved by the elimination of the social basis of classes.

> (11) The philosophers have merely *interpreted* the world differently; everything depends, however, on *transforming* it.

Mere interpretation is not to be replaced by mere activity; the two moments gain new meaning from a pervasive unity. The initial phase of demythologizing is motivated by a critical suspicion that the best of all possible worlds is a bitter struggle away, a suspicion which seeks a different interpretation of the world in the hopes that its specifics will point the way to a social transformation and help rally the necessary forces. The initial step of interpretation is therefore crucial. From it must follow the self-education of the revolutionaries: a theory of the society to be transformed: ideological weapons for the causes and a strategy based upon immanent potentials of both the process of change and the establishment of a better world. To accomplish these tasks coherently and powerfully, the ideology critique cannot be arbitrary. Critical thought must burrow into the core of the social world, contemporary practice, the production of daily social existence, the creation of commodities and the reinforcement of the commodity mode of production.

The unity of theory and practice in Marx's work is not the subjugation of thought to political activism. Practice involves above all the production of material goods, but also the reproduction of social relations (including the relations of production), attitudes, legal order, military defense, etc. *Capital*, a theoretical work, stands in a unity with the practice of nineteenth century European society in a number or ways:

(a) It is a reflection upon that practice, bringing people to a comprehension of the practice in which they are involved.

(b) Thereby, it is part of that practice itself; the self-conscious moment.

(c) It is conditioned by that practice, which provides its motivation and possibility in many ways.

(d) Its goal is to restructure that practice by pointing to structural contradictions and potentials for change.

(e) In these mediated, namely theoretical, ways, *Capital* is a political act.

By focusing on the realm of production, Marx was able to concretize each of these points of unity in terms of the proletariat, thereby arriving at a concrete theory of the potential for transforming the world in terms of a revolutionary subject-object of theoretically informed transformative practice.

Marx believed that every social theory is conditioned by its social context, so that for every interpretation of the world there is a relation and interpenetration of theory and practice. Marx's own approach differs from others – whether contractual, idealistic or scientistic – in that his consciousness of this relation led him to *unify* his theory and social practice by constructing his methodology primarily in terms of this relation. Recognizing that his manner of abstraction and his central categories had their conditions of possibility in the society whose structure they were designed to articulate, Marx could clearly define the socio-historical specificity of his concepts and he could perceive their interrelationships in terms of the structure of bourgeois society. Above all, Marx's critical thrust was a conscious response to the social mystification arising from the fetishism of commodities; it is as a conscious response to the ideological character of capitalist society that Marx transforms philosophy into ideology critique and the critique of political economy. The following chapters explore the result of this transformation of philosophical interpretation of the world into transformative interpretation, the critical hermeneutic of capitalist society.

Chapter V. Research: *The Grundrisse*

The first two dozen pages of the *Grundrisse*, the notes and rough drafts of Marx's economic studies, which were later published within a more restricted scope in *Capital*, provide an extended treatment of Marx's method. Valuable discussions of the following three issues can be found in this "Introduction," which guided Marx's economic research: (1) the centrality of commodity production, (2) the method of analyzing capitalist society, and (3) the relationship of transhistorically general to socio-historically specific concepts.

The important chapter of Marx's *Grundrisse* entitled "Forms which Precede Capitalist Production. (Concerning the Process which Precedes the Formation of the Capitalist Relations of Original Accumulation)" elaborates upon several themes presented in the "Introduction." The forty pages of this chapter contain the only extended consideration of pre-capitalist societies in the thousand pages of analysis, yet here Marx actually becomes repetitive. In the formulations of his reconstructive approach to history and in his historical characterization of the notion of property, so important are they to his concerns. Marx's reconstruction of the history of property relations, understood in their relation to the prevailing mode or production provides a unity to Marx's doctrines, presents the core of his historical materialism, and supplies the critical fulcrum for his critique of political economy.

Materialistic Conceptualizations for the Self-Interpretation of the World

Marx's statement of the priority of the category of production within an economic analysis of bourgeois society is so straightforward that it requires little comment. He begins his analysis of the general relations of production to distribution, exchange, and consumption by giving the traditional definitions of these terms as used by bourgeois economists (Smith, Ricardo, Mill, etc.):

The obvious notion: in *production* the members of society appropriate (create, shape) the products of nature in accord with human needs; *distribution* determines the proportion in which the individual shares in the product; *exchange* delivers the particular products into which the individual desires to convert the portion. which distribution has assigned to him; and finally, in *consumption*, the products become objects of gratification, of individual appropriation. . . . Thus *production* appears as the point of departure, *consumption* as the conclusion, *distribution* and *exchange* as the middle, which is, however, itself twofold, since distribution is determined by society and exchange by individuals.[41]

The four basic spheres of the capitalist economy are posited by bourgeois political economy as autonomous domains to be found in every economic system. Marx's analysis is dialectical, negating the theoretical limitations involved in viewing these domains as autonomous by explicating their conceptual, structural and historical interdependence, finally arriving at production as in each case the historically determinant and determining essence behind the multiple appearances. Appealing to the traditional definitions just stated (e.g., consumption as appropriation of the object by the individual). Marx develops the relation of each domain to production in a strikingly Hegelian style. Concerning production and consumption, he thus argues:

Production is also immediately consumption. Twofold consumption subjective and objective: the individual not only develops his abilities in production, but also expands them, uses them up in the act of production, just as natural procreation is a consumption of life forces. Secondly: consumption of the means of production, which become worn out through use, and are partly (e.g., in combustion) dissolved into their elements again. Likewise, consumption of the raw material, which loses its natural form and composition by being used up. The act of production is therefore in all its moments an act of consumption. . . . The

[41] Karl Marx, "Introduction," *Grundrisse* (London: Penguin, 1973), p. 88f, emphasis added. Cf. Karl Marx, "Einleitung," *Grundrisse der Kritik der politischen Ökonomie (Rohentwurf) 1857-1858* (Frankfurt: Europäische Verlagsanstalt, photocopy of 1939, 1941 Moscow editions), S. 10f. The introduction was the only section of the *Grundrisse* published before 1939; An alternative, but inferior translation can be found as an appendix to Karl Marx, *A Contribution to the Critique of Political Economy* (New York: International, 1970), p. 188ff.

artistic object – like every other product – creates a public which is sensitive to art and enjoys beauty. Production thus not only creates an object for the subject, but also a subject for the object. . . . The important thing to recognize here is only that, whether production and consumption are viewed as the activity of one or of many individuals, they appear in any case as moments of one process, in which production is the real point of departure and hence also the predominant moment.[42]

In the case of production for immediate consumption, the economist's "state of nature," production is decisive. However, commodity production is by definition mediated by distribution. Marx therefore turns next to the relation of this latter sphere to production, in order to see what changes this added complication may introduce to the centrality of production:

An individual whose participation in production takes the form of *wage labor* will receive a share in the products, the results of production, in the form of *wages*. The structure of distribution is entirely determined by the structure of production. Distribution itself is a product of production, not only in its object, in that only the results of production can be distributed, but also in its form, in that *the specific kind of participation in production determines the specific forms of distribution*, i.e., the pattern of participation in distribution. It is altogether an illusion to posit land in production, ground rent in distribution, etc. . . . In the shallowest conception, distribution appears as the distribution of products, and hence as furthest removed from and quasi-independent of production. But before distribution can be distribution of products, it is (1) the distribution of the members of society among the different kinds of production. (Subsumption of the individuals under specific relations of production.) The distribution of products is evidently only a result of this distribution, which is comprised within the process of production itself and determines the structure of production. . . . The question of the relation between this production determining distribution, and production, belongs evidently within production itself. If it is said that, since production must begin with a certain distribution of the instruments of production, it follows that distribution at least in this sense precedes and forms the presupposition of

---

[42] *Ibid.*, p. 90-94, S. 11-15.

> production, then the reply must be that production does indeed have its determinants and preconditions which form its moments. At the very beginning these may appear as spontaneous, natural. But by the process of production itself they are transformed from natural into historic determinants, and if they appear to one epoch as natural presuppositions of production, they were its historic product for another.[43]

Working with an initially "ahistorical" concept of distribution, that is, a concept which had not previously been subjected to historical reflection, Marx points to its role in the historic process and shows how its form is determined by its socio-historical position. In capitalist society, for instance, distribution takes the form of wages and profit – a mode of distribution which is determined in its general form and specific contents by the relations of production, the social division of labor into alienated wage labor and private ownership of the means of production.

Exchange, one of whose phases is circulation, is itself simply an intermediary phase between production and distribution and consumption. As these latter domains are essentially determined by production in its broader sense, exchange is easily seen to stand in a similar relation to production:

> Exchange appears as independent of and indifferent to production only in the final phase where the product is exchanged directly for consumption. But (1) there is no exchange without division of labor, whether the latter is spontaneous, natural, or already a product of historic development, (2) private property presupposes private production, (3) the intensity of exchange, as well as its extension and its manner, are determined by the development and structure of production. For example, exchange between town and country; exchange in the country, in the town, etc. Exchange in all its moments thus appears as either directly comprised in production or determined by it.[44]

Marx accordingly concludes this section of his "Introduction" with a statement of the priority of production:

> A definite production thus determines a definite consumption, distribution and exchange as well as *definite*

---

[43] *Ibid.*, p. 95-97, S. 16-19; emphasis added.

[44] *Ibid.*, p. 99, S. 20.

relations between these different moments. Admittedly, however, *in its one-sided form,* production is itself determined by the other moments.[45]

A number of comments are appropriate here:

(a) Marx's conclusion about production is of extreme methodological import. An analysis of a given economic system must, according to the argument, begin in the realm of production, proceeding to an analysis of the remaining economic domains only on the basis of insights gained from an understanding of the mode of production (e.g., production of tribal value, of immediate use value, or, in capitalism, of commodities). Any other analytic approach cannot capture the uniqueness of the economic categories in the given structure because their specificity is founded in the specific mode of production. An analysis of capitalism that is interested in uncovering the differences from previous and from potentially future economic systems must begin with an analysis of commodity production.

(b) Marx has distinguished "specific" concepts from "general." The definitions of production, distribution, exchange and consumption that Marx borrowed from traditional political economists as his starting point represent general concepts, presumed applicable to the description of any system of social institutions which responds to the human condition. Specific concepts, by contrast, state the difference between, for instance, feudal distribution (tribute, alms, etc.) and capitalist distribution (wages, rent, interest). Each of these specific concepts points back to the corresponding specific concept of production: feudal agricultural production or capitalist commodity production. (Cf. the following section.)

(c) All roads led to Rome in the Roman Empire and all approaches to the capitalist system lead inevitably to capitalist production, provided only that one perseveres. Of course, not all empires radiate out from as absolute a center as Rome, and not all social systems are as tightly integrated and dependent upon the realm of production as those with capitalist relations. Only in capitalist society does the distribution of goods follow from the role in production according to strict mathematical calculation (wages based upon labor time). Previously, hereditary ties, social hierarchy and a host of other non-rational considerations mediated the relationship of distribution to production. For the future, too, Marx hopes for a society in which there is a relative autonomy for each domain, so that each could respond to the criteria peculiar to it alone, escaping from restrictions imposed by the mode of production. The communist slogan, from each according to his ability; to each according to his needs," formulates precisely the non-identity which

---

[45] *Ibid.*

transcends the inhuman limitations of capitalism. Marx's argument for the centrality of production in economic relations is thus of special relevance to a theory of capitalist society, a social system which ruthlessly imposes economic categories on all realms of life. In rejecting the ahistorical character of economic concepts, Marx consciously situates his theory – inclusive of its philosophical foundations – in its historical setting. In capitalism, the productive sphere attains a clear priority. Economics rules all spheres of society. Labor as free labor becomes labor-as-such, "abstract labor." Thereby production, as commodity production, becomes production-as-such, abstract production. Retrospectively, we can then view human labor as the universal source of creation. Similarly, production – always tautologically a precondition of social existence – can then retrospectively be viewed as the essential realm for the analysis of every society. Thereby, the dialectic of forces and relations of production – which is always part of social mediations but which has its paradigm instance in the industrial revolution where it constitutes the essence of the social development – can also be extended retrospectively throughout history. Capitalist social structures are a key to less developed ones in which forces other than production (religion, politics, etc.) may have obscured the role of production. The point of such retrospective views is primarily to shed light on capitalist structures in terms of their prehistory, rather than to understand previous societies on their own terms. Within this project the retrospective universality is undogmatic. (This argument will be substantiated in the final section of Chapter V in terms of Marx's texts.)

Marx's is an essentially dialectical and historical science. Despite important differences with them, Marx clearly chose Hegel and Darwin as partial models of scientific method. In his "Introduction" to the *Grundrisse* he both elaborates upon Hegel's opposition of the form of *research* to that of *presentation* and uses the notion that "the anatomy of man is a key to the anatomy of the ape."

Hegel's distinction has been reviewed in terms of a discussion in Hegel's *History of Philosophy* and related to Marx's method elsewhere.[46] The distinction – foreshadowed at least by Kant in his own way in the contrast of the "regressive" approach of the *Prolegomena* or *Foundations* compared to the "synthetic" approach of the first and second *Critiques* – can also be seen at work in the relation of Hegel's *Phenomenology of Mind* to his *Science of Logic*.[47] One can, that is, view the *Phenomenology* as the record of a research which

---

[46] Alfred Schmidt, *Geschichte und Struktur* (München: Hanser, 1971), S. 52ff.

[47] Cf. G. W. F. Hegel, "With what must science begin?" and "Introduction," both in his *Science of Logic*.

traces the development of mind quasi-historically from its beginnings in sense-perception to its culmination in absolute spirit. The *Science of Logic* starts then from Being as the (sublimated) conceptual representation of absolute spirit and proceeds to unfold this concept systematically to account for all categories of mind. Whereas the *Phenomenology*, so viewed, would illustrate the form of *research*, paralleling a more commonsensical or naturally historical progression, the *Logic* would illustrate the form of *presentation*, proceeding from a central concept and logically deriving the entire science from it. Rejecting Hegel's emphasis on the realm of mind, Marx retains Hegel's critique of empiricism, of the view that science consists in subsuming cases under classifications that have merely pragmatic significance. The point for Marx is to get at the essence of the matter (capitalism) by a process of research that begins with the obvious appearances, but then to present the matter systematically, comprehended in terms of its essence. We have already seen this notion of comprehension at work in the manuscript on alienated labor, where Marx criticized political economy for not grasping the essential connections between their concepts. Later, we shall see how *Capital* embodies Marx's conception of a dialectical science of society. First, however, we must look at Marx's development of this ideal.

Marx begins by criticizing the common-sense assumption that "it seems to be correct to start with the real and the concrete, with the real precondition, thus to begin, in economics with e.g., the population, which is the foundation and the subject of the entire social act of production."[48] Marx rejects population as a concrete element, claiming that it is in fact an abstract notion, a representation which overlooks the essential features of population, such as class structure. Population cannot be used as the starting point for an interpretive theory of society, for its concreteness is merely apparent simplistic, hence false. The notion of population may, however, suggest itself as the starting point of an investigation leading to the interpretive essence: "If I were to begin with population, it would be a chaotic representation of the whole and through closer determination I would arrive analytically at increasingly simple concepts; from the represented concrete to thinner and thinner abstractions until I reached the simplest determinations."[49] One must, that is, in Walter Benjamin's phrase, "transverse the icy wasteland of abstraction in order to arrive conclusively at concrete philosophizing."[50]

The investigation of population in terms of classes and of these in terms of the factors which define them – wage labor, capital, commodities, surplus

---

[48] "Introduction," *Grundrisse*, p. 100, S. 21.

[49] *Ibid.*, p. 100, S. 21.

[50] Theodor W. Adorno, *Negative Dialektik* (Frankfurt: Suhrkamp, 1966), S. 7.

value, etc. – leads to concepts like exchange value, which are theoretical constructs abstracted from the social process. Using such conceptually simple terms, one can then analyze complex phenomena coherently and systematically. The research moves from appearance to essence, from the imaginary concrete to the abstractly simple, and then back again: "From there it would be necessary to make the journey again in the opposite direction until I had finally arrived once more at the population, but this time not as the chaotic representation of a whole, but as a rich totality of many determinations and restrictions."[51] The return journey is the systematic presentation. It moves from the "simple," "abstract," essential, theoretical concepts such as value back to the complex phenomena such as population – now replacing the seemingly concrete rational mass with a truly concrete totality, conceptually analyzed as, say, a system of classes, understood in terms of their positions in the process of production.

The image of the round-trip journey can help us follow Marx's progress beyond bourgeois political economy, as he viewed it, recognizing, of course, that this attempt to apply the distinction of research and presentation is grossly simplistic and can hope only to be suggestive. The trip begins with "vulgar" political economy, those writings which accepted the necessary illusions projected by the capitalist system and propagated them to ideological ends. Based on the realm of exchange, these theories used the ideology of just exchange to claim that capitalism was natural, rational and the final stage of history. "Classical" political economy (Petty, Smith, Ricardo, etc.) made tremendous strides along the research route, reaching the insight into the centrality of production (labor) and formulating a labor theory of value. In his own economic studies, preserved primarily in the *Grundrisse* and the *Theories of Surplus Value*, Marx followed the progress of the classical theorists, clarifying metaphysical confusions, developing his own accounts of the commodity and surplus value and, above all, placing the whole in a radically historical context. Then in *Capital*, the explicitly scientific return trip, Marx started from the abstract simples, commodity and surplus value, to develop a systematic account of capitalism, beginning in Volume I with the essential realm of production.

The continuous dialectical motion of anticipation, research and presentation or of vague image, abstract concept and concrete articulation is, in the first approximation, very common. It involves tentatively constructing an interpretation of some subject matter while investigating it, later revising the interpretation in light of new findings. First impressions on meeting someone are an instance of projecting a personality structure onto a set of behavior,

[51] "Introduction," *Grundrisse*, p. 100, S. 21.

allowing for subsequent reinterpretation on the basis of increased familiarity. Heidegger has conceptualized this in connection with the problem of becoming familiar with Being in terms of pre-understanding, the fore-structure of understanding, the hermeneutic circle, and the relationship of *Being and Time* to his later works. If Heidegger were now to give a presentation as systematic as *Being and Time* of his research – the questioning of Being – many elements of his various studies would be present but false leads and faulty aspects would be suppressed as irrelevant (except insofar as their failures had an insightful necessity). So, too, in Marx's case, many concepts and considerations of the early inquiries, critiques and self-clarifications appear in *Capital* – so much so that commentators often claim that Marx's mature outlook or even his economic system were already worked out in the very early manuscripts. However, in the final presentation many of the accidental emphases stemming from Marx's biography rather than from the nature of capitalism have been dropped, details filled in, insights deepened, concepts greatly developed, interconnections systematically drawn, and a coherence impressed upon the whole.

The *process* of theory-building allows the presentation to be appropriate to the object of the research and the abstract concept to be appropriate to the articulation of the concrete phenomenon. Although there is a unity of preconception, analysis and systematization, the mediating conceptual framework is not determined in advance. Marx began with the concepts which had already grown out of political economic research. As self-articulations of bourgeois society, these concepts were both suggestions for an appropriate conceptual scheme and symptoms of the self-mystifying power of commodity relations. Marx criticized the deviations of the concepts from the present economic reality and greatly developed their specificity, interconnections and content in the course of his own research. Not predetermined, the conceptual framework does not come only at the end of the research, through induction. The research cannot proceed without concepts, but is itself the process of searching for, testing and criticizing interpretive concepts. Only the concepts can determine what counts as a fact, as a concrete social phenomenon. Empirically observable and statistically calculable facts *have already been mediated in reality by social forces* and only their *mediation by abstract concepts in reflection* (theory, analysis) can uncover this, their essential character. Concepts and facts cannot be thought without each other. Each abstraction must be developed in the process of research, a process which can only thereby lead to a more complete, concrete grasp of its object. The systematic presentation then consists merely of a retrospective view of the research, organized in accordance with the results of the research. The dialectical unity of presentation and research assures that the interpretive system treats its object as unique, as socio-historically specific.

Because Marx resists imposing an *a priori* system upon his research, but develops his analytic terms out of the 'thing itself' as it presents itself in the research, Marx's method must be considered *more hermeneutical than metaphysical*. Hermeneutics, as the art of interpretation, characteristically comes to the fore when interpretation has become problematic, and this is certainly the case with Marx's object. The fetishism of commodities has obscured the social relations definitive of bourgeois society, turning naive theories of society into socially necessary illusion: ideology. Marx's task is thus typical for the hermeneute, namely to interpret the given *as* false or distorted solely on the basis of that given itself, without imposing an external interpretive scheme. The relationship of his research to his presentation is one aspect of Marx's response to this task.

Research into bourgeois society as a whole should, if allowed to develop fully, eventually come upon the concepts of commodity, value, etc. regardless of its starting point. Scientific *analysis* of specific phenomena must, on the other hand, begin with the proper concepts to reach an understanding of the phenomena within the structure of society. The essential concepts depend on the form of society, on the form of production which stands at the center of society.

Marx's position is thus strictly opposed to an ahistorical approach which seeks to deduce the character of the present from an absolute origin. He is opposed to the idealist goal of an unconditioned *prima philosophia* which proceeds from an indubitable truth like Descartes' *cogito sum* or Hegel's sense-certainty and immediate Being. The essence of the present is to be found in the present, not in some imaginary or even real past where that essence may have existed in some undeveloped and peripheral form.

> For instance, nothing seems more natural than to begin with ground rent, with landed property, since this is bound up with the earth, the source of all production and of all being, and with the first form of production of all more or less settled societies – agriculture. But nothing would be more mistaken. In all forms of society there is a specific branch of production which determines the stature and influence of all the other branches, whose relations do this for the relations of all the other branches. There is a pervasive illumination in which all other colors are bathed and which modifies them in their specifics.[52]

In agricultural society, for instance, the crafts and industries that are present in a crude state adopt the characteristics of agricultural relations: the

---

[52] *Ibid.*, p. 106f, S. 27.

immediate relationship to specific human needs; organization of property, knowledge and roles along hereditary lines; and so forth. In capitalist society, the opposite takes place: even agriculture becomes a commodity industry.

Just as delimited analyses of particular social phenomena in the capitalist era must rely upon the central categories of capitalist production, so a presentation of the whole social system must begin with them. It would therefore be unfeasible and wrong to let the economic categories follow one another in the same sequence as that in which they were historically decisive. Their sequence is determined, rather, by their relation to one another in modern bourgeois society, which is precisely the opposite of that which seems to be their natural order or which corresponds to historical development.[53]

The economic categories are thus interrelated in several ways: in terms of the existing social system, according to their historical development and in a natural or logical order. Darwin's theory of the origin of man provides an analogy to this situation. Suppose one asked about man's prehensile thumb. The naively historical analysis might point out that first fish developed protruding gills, then mammals acquired limbs with fingered hands and finally the primate family evolved the opposable thumb, which we have inherited. Someone with more awareness of epistemological issues could, like Marx, add a twist: "The intimations of higher development among the subordinate animal species, however, can be understood only after the higher development is already known."[54] The claim that modern man's thumb is the *telos* of the prehistoric gill is not based on religious faith, but on a perspective shared by the Hegelian owl of Minerva: the present. In taking human anatomy as the key to that of the ape – bourgeois economy as the key to the economy of antiquity – Marx's historical approach is radically *reconstructive*, as opposed to naively constructive historiology. Marx carries out his reconstruction of pre-capitalist economic formations on the basis of his knowledge of their *telos* as now actualized in bourgeois society. (Cf. the section on Marx's retrospective history of property relations.)

By deriving the succession of his categories from the relation they have in bourgeois society, Marx avoids the traditional dilemma of philosophy up to Hegel. Hegel, that is, had already made theory and history relative to each other in dialectical fashion, but for him this meant either filling the logical sequence of the categories with a wealth of historical material or else rationalizing (sublimating, abstracting) actual history into the shape of a sequence of forms. Marx's radical unity of theory and history has no need for

---

[53] *Ibid.*, p. 107, S. 28.

[54] *Ibid.*, p. 105, S. 26.

these imposed techniques. The critique of political economy, as the critical theory of bourgeois society, abolishes the duality of philosophy and historical analysis by discovering the entire history of the categories and forms of social existence in present society. Conversely, as critical, Marx's theory rejects the facade of immediacy in given appearances by presenting their concrete historical genesis. It dissolves or "de-constructs" the mystifying process of reification and fetishization by inquiring after the social and historical conditions and presuppositions of present-day appearances and concepts. If Marx's concepts are reflections of bourgeois society, they have also been critically transformed in the recognition that the conceptual mirror is systematically warped by the fetishism of commodities. Through this form of appropriation of the given categories – determined by socio-historically specific considerations already based on his theory – Marx avoids the ideological thrust of those theories which lack such a conscious unity of theory and practice.

Marx's historical method of analysis had important consequences for his manner of forming concepts as well. Through the historical character of his conceptualizations, Marx avoided the problems created by the "vulgar" and the "classical" political economists due to their unreflected perspective on history. While Marx agrees that previous social forms are to be understood from the perspective of bourgeois society – viewing history as leading up to capitalism and using the economic concepts of capitalism as analytic tools – he insists that the differences between that which the concepts denoted then and now not be over-looked. It is, he thinks, quite impossible to use the present system as a key to preceding formations if one tries to do this:

> in the manner of those economists who obliterate all historical differences and who see in all forms of society only bourgeois relations. If one is acquainted with ground rent, one can understand tribute, tithe, etc. But they do not have to be treated as identical. . . . Thus, although it is true that the categories of bourgeois economics possess a truth for all other forms of society, this has to be taken *cum grano salis*. They may contain them in a developed, stunted or caricatured, etc., form, but always with an essential difference.[55]

The general concept, rent, is valid for all or at least many social systems, but, on the other hand, the specific concept of bourgeois rent is essentially different from feudal rent (tithe) or classical rent (tribute). The common confusion among political economists between the generality and specificity

---

[55] *Ibid.*, p. 105f, S. 26.

of their concepts deserves extended consideration and the present text is perhaps the most explicit of Marx's writings on this point. (Cf. the following section.)

Marx's thought is historical in yet another way. In addition to viewing social phenomena as developments in a retrospectively historical manner and clarifying the socio-historical specificity of his concepts, Marx relates his work to the present state of society, to the strivings of social groups. Highly conscious of the fact that his theory is situated in the bourgeois era, Marx does not make a fetish of capitalism as though it were necessarily the final stage of historical progress. Clearly, his intent is quite the opposite, and the subjective intent is objectively embodied in the theory's structure, providing its unity of theoretical and political practice. Further, the possibility of embedding his revolutionary intent in economic theory has its foundation in the reality of the class struggle:

> The so-called historical presentation of development is founded, as a rule, on the fact that the latest form regards the previous ones as steps leading up to itself, and, since it is only rarely and only under quite special conditions able to criticize itself – leaving aside, of course, the historical periods which appear to themselves as times of decadence – it always conceives them one-sided.[56]

The important and intriguing "Afterword" (1873) to the second German edition of *Capital* clearly demonstrates how important the relationship of the social context (working class consciousness, academic theories and ideologies, economic crises) to *Capital* was for its author. Marx sees the revolutionary character of his dialectical method in the fact that it treats the present stage as not only the progressive culmination of the past, but also as a contradictory and repressive system which is, however, in the process of transcending itself towards the future:

> In its rational form dialectics is a scandal and abomination to bourgeoisdom and its doctrinaire professors because it includes in its comprehension an affirmative recognition of the negation of that state, of its inevitable breaking up; because it regards every historically developed social form as in fluid motion, and therefore takes into account its transitory nature not less than its momentary existence; because it lets nothing impose upon it, and it is in its essence critical and revolutionary.[57]

---

[56] *Ibid.*, p. 106, S. 26.

Historically-specific Conceptualizations

Production is always production of a particular product, by a particular worker, with particular means in a particular social context; the concept of production must take this into account. Not one to pull his conceptual punches, Marx opens the "Introduction" to his critique of political economy in the *Grundrisse* with an immediate consideration of this aspect of his central category: production. His first line makes clear that he uses this term in an historically or socially specific sense; "material production" refers to productive forms specific to a particular historical period or to particular branches of production, not to production in general or general production: "The object before us is, to begin with, *material production*. Individuals producing in society – hence socially determined production of individuals – is, of course, the point of departure."[58] The production which keeps society going is production within society, determined by the character of the particular society, carried out by agents acting in socially defined roles.

What of someone who today produces, say, pottery the way it has been produced by hand in the most varied societies? Firstly, it is undoubtedly done as a hobby or art form, outside the concerns of society and primarily for personal enjoyment and expression. Further, it most likely uses modern technology: chemically pure glazes or an electric kiln – especially if there is some hope of competing with factory-produced ceramics. And clearly the aesthetic sensibility at work will be in the end peculiarly modern (especially, again, if there is an appeal to a buying public), perhaps a reaction against the consequences of mass production.

Additionally, Marx argues, the very possibility of producing as an individual – developing a superfluous hobby outside of the socially efficient forms of production – is itself a product of history, of social and productive development to the point where workers have free time. Marx takes the novel about Robinson Crusoe, a totally eighteenth-century British gentleman parading as nature boy, as a typical example of the ideology of individualism. He points out that the individual, both as a category and as a mode of existence, is peculiarly bourgeois, a result of the historical developments leading to capitalism. Against the bourgeois projection of the concept of the individual back to the nomadic or tribal beginnings of history, Marx notes that the biological human entity was part of a social grouping or social formation out of which and on whose basis alone it could define itself

[57] Karl Marx, *Capital*, Vol. I (New York: International, 1967), p. 20. Cf. Karl Marx, *Das Kapital*, Bd. I (Frankfurt: Ullstein, 1971), S. 12.

[58] "Introduction," *Grundrisse*, p. 83, S. 5.

individually. The argument extends Hegel's dialectic of mutual recognition and anticipates Wittgenstein's rejection of private languages, stressing that individualism is essentially (biologically and historically) social, not "natural":

> The more deeply we go back into history, the more does the individual, and hence also the producing individual, appear as dependent, as belonging to a greater whole; in a still quite natural way in the family and in the family expanded into the clan; then later in the various forms of communal society arising out of the antitheses and fusions of the clans. Only in the eighteenth century, in bourgeois society, do the various forms of social connectedness confront the individual as a mere means towards his private purposes, as external necessity. But the epoch which produces this standpoint, that of the isolated individual, is also precisely that of the hitherto most developed social (from this standpoint, general) relations. The human being is in the most literal sense a *zoon politikon*, a political animal, not merely a gregarious animal, but an animal which can individuate itself only in the midst of society. Production by an isolated individual outside society – a rare exception which may well occur when a civilized person in whom the social forces are already dynamically present is cast by accident into the wilderness – is as much of an absurdity as is the development of language without individuals living *together* and talking to one another.[59]

Consequently, the two ahistorical approaches to defining production are equally unacceptable; the proper approach lies between the extremes. The idealist definition of production as the forming of the object by a subject with the aid of a mediating tool completely misses the particularity of object, subject and tool as belonging to an historical stage and a social system. The nominalist definition, on the other hand, misses the common element of instances of production within a given socio-historical context as well as obscuring the underlying fact that subject, object and tool are already results of historical and social processes. Production as the mediation of individual subject and natural object is itself as much mediated by society as are its two poles. Production takes place within one social context or another and is essentially defined by the society's specifics. Even the rare cases of individuals producing outside the social norms (castaways, hermits, artists) are merely exceptions that prove the rule in that their deviations presuppose the norm.

---

[59] *Ibid.*, p. 84, S. 6.

The conclusion for a theory of capitalist society is that the concept of production must always be in terms of the specific concepts of production appropriate to specific societies, e.g., commodity production in bourgeois society. Marx draws this conclusion, excluding the notion of the conceptual priority of production-in-general:

> Whenever we speak of production, then, we always have in mind production at a definite stage of social development – production by individuals in a society. It might therefore seem that, in order to speak of production at all we must either trace the various phases in the historical process of development or else declare beforehand that we are dealing with a specific historic epoch such as, e.g., modern bourgeois production, which is indeed our particular theme.[60]

Marx does leave room for the concept of production-in-general – neither socially nor historically specific – but his way of doing this shows why the *particular* concepts of production have a priority, rather than legitimating the liberal ideology which exploits general concepts as ahistorical. His characterization of the universal genus, similar to Wittgenstein's notion of family resemblance, seems strikingly positivistic.

> All periods of production have, however, certain common traits, common characteristics. *Production in general* is an abstraction, but a rational abstraction in so far as it really brings out and fixes the common element and thus avoids repetition. . . . Some determinations belong to all epochs, others only to a few. The most modern epoch and the most ancient will share (some) determinations.[61]

Of course, Marx is by no means a positivist: he is merely choosing his level of abstraction on the basis of the nature of his subject matter, rejecting the extremes of ahistorical abstraction and positivistic data collection, but synthesizing what is valuable or critical in each of the approaches by developing his theory with abstractions which are filled with content: specific concepts. Categories like commodity production allow Marx to work out a dialectically scientific theory of the essential structure of the capitalist system, while simultaneously allowing him to distinguish what is necessary to the reproduction of mankind in its dialectic with nature and what is merely imposed. Such categories are crucial to practical philosophy, to transformative interpretation, to critical theory.

---

[60] *Ibid.*, p. 85, S. 6f.

[61] *Ibid.*, p. 85, S. 7.

A *general* concept such as production in general, as Marx uses it, is different from the *ahistorical* concepts of bourgeois political economy in that it is the result of historical reflection, a synthesis of *specific* concepts of production or a summary of what is common to the specific concepts. In this form, general concepts play an important role in Marx's theory – not on their own as metaphysical preconceptions, but in the contrast with their specific sub-concepts. Of course, this is strictly speaking true only in *Capital*, Marx's developed and systematic presentation. In his early manuscripts, terms occasionally appear in an ahistorical form anticipating the general concepts which will subsequently result from historical analysis, a more highly developed systematic context and a more extensive critique. However, even these early occurrences are results of specific negation of idealistic or ideological outlooks in terms of social and historical considerations, and thus Marx's usages are less ahistorical than they might appear.

Marx uses one of his favorite analogies, that of language to production, to point out the critical necessity of contrasting the specific with the general concept. He notes that although the most highly developed languages have laws and characteristics in common with the least developed languages, it is precisely their divergences from these general and common elements which determine their development.[62] Language, as a medium for interpersonal communication, has certain necessary elements which all languages have shared. But there is a second set of elements to be discovered in any actual language which are not necessary to the task of human communication and which may actually restrict the freedom of expression. Progress in the evolution of language, then, consists partially in rejecting these restrictive features.[63]

The analogy applies directly to the specific concepts of production, so that the socio-historically specific elements "must be separated out from the determinations valid for production as such, so that in their unity – which arises already from the identity of the subject, humanity, and the object, nature – their essential difference is not forgotten."[64] The notion of "second nature" is helpful here; it refers to conditions or institutions which appear natural, necessary, inevitable, but which are in fact created by people,

---

[62] *Ibid.*

[63] Marx's remarks on language have recently been developed by Apel and Habermas, whose theory of communicative competence tries to define those necessary features of language in order to distinguish them from the "systematic distortions" which may ideologically restrict the attempt to use communication to achieve emancipation from social relations of domination.

[64] *Ibid.*

contingent, not necessary. Private property is, for instance, a legal institution which has become second nature to us, but which, Marx argues, was – in distinction to possession – absent from primitive societies and now acts as a chain upon us, keeping us from a better society in which it would disappear. The general concept of production, then, would state what is natural (necessary) to production in all ages and all social systems. The difference between the specific concept of bourgeois production and this general concept of production would define bourgeois second nature. The ideology of the capitalist era is largely based on the positing of second nature as natural, on assuming that the characteristics peculiar to the bourgeois age are eternal, rational and sacred. Only a social theory which distinguishes necessity from second nature and explores the possibilities for eliminating the latter can perform the critical function rather than one of blanket justification.

While the notion of necessity has here been introduced in an ahistorical manner, Marx's notion of the general concept retains historical content; while the necessary nature of production can never be diverged from, Marx's "production-in-general" merely has not yet been diverged from. Marx's parenthetical remark that the generality can be understood by merely noting that production always involves humanity and nature, thus retaining a minimal invariant content, suggests an ahistorical side-glance at the necessary structure. But his remark has the character of a tautological aside and he does retain his historical definition of general concepts throughout. The ahistorical, almost tautological side-glance appears as a moment of abstraction from history within a discussion focused on history. Marx is well aware that neither the subject (the worker) nor the object (nature) of the production process remains invariant through the history of the modes of production. But the abstraction which allows us to identity the poles of production with the terms "humanity" and "nature" also allows us briefly to posit the unity of production-as-such – in order thereby to determine the historical *differentia specifica* of particular modes of production.

Marx's revolutionary anticipations of a society without certain elements of bourgeois second nature do not, therefore, rest on speculation about the true nature or essence or human production as on a metaphysical first principle, but on an historical argument, namely that those institutions which are second nature to us now did not exist at certain periods in human history and therefore are not necessary to human life and may conceivably be eliminated in the future. This argument certainly does not in itself prove that alienated labor or private property will inevitably disappear or even that the benefits of capitalism could be maintained without the undesirable aspects of capitalist second nature. It is primarily an argument against the outlook of bourgeois political economy which took features peculiar to capitalism as unquestionably necessary for all societies. Marx's thought is thus historical not only in the sense that his analytic standpoint is consciously rooted in the present, but also in that his hope for the future has its basis in unachieved potentials of the past and present.

General concepts are historical in another, more striking way than merely as summaries of historical instances. The supposedly ahistorical concepts of bourgeois political economy – production, labor, property, value, etc. – are, according to Marx, merely the sublimated expression of the respective specifically bourgeois concepts – commodity production, wage labor, private property, labor value, etc. These specific concepts are what Marx calls simple categories, as opposed to what is naively imagined to be concrete, like national population.

Simple categories are historically situated: they presuppose concrete historical developments for their very meaning. "For example, the simplest economic category, e.g., exchange value, presupposes population, a population moreover which produces in specific relations, as well as a distinct kind of family or community or state, etc. Exchange value cannot exist except as an abstract, one-sided relation within an already given, concrete, living whole."[65] Marx's materialism views the concrete social totality, which is the result of comprehending the imaginary concrete with abstract simple categories, as "a product of the working-up of observation and representation into concepts."[66] A certain historical stage must already have been reached and perceived before the image of it could have been transformed into the categories which are appropriate to its comprehension. Theory does not approach a social formation with concepts created through unsituated contemplation: the theory of capitalism, down to its simplest and broadest categories, presupposes the historical development of capitalism: "Hence, in the theoretical method, too, the subject, society, must always be kept in mind as the presupposition."[67]

The clearest example of the relation of a general to a specific category is that of labor. The attempt by Marx's critics to saddle him with an ontology of labor or a metaphysics of *homo faber* relies upon ignoring the distinctions among the ideologically ahistorical, historically general and specifically bourgeois categories of labor. The particular category of bourgeois labor was formulated by Adam Smith, who "rejected every limiting specification of wealth – productive activity – not only manufacturing, or commercial, or agricultural labor, but one as well as the others, labor in general."[68] The

---

[65] *Ibid.*, p. 101, S. 22.

[66] *Ibid.*

[67] *Ibid.*, p. 102, S. 22.

concept of labor-as-such, "abstract labor," presupposed the social system which was in fact establishing itself in Adam Smith's day. In capitalism all labor becomes equivalent, measured quantitatively by labor-time, regardless of qualitative differences. The bourgeois job and labor markets ironed out all differences between particular forms of labor or types of workers, transforming labors into labor as such. Smith's concept expressed his perception of this social development. The fact that the particular type of labor is irrelevant both socially and conceptually "presupposes a highly developed totality of real kinds of labor, of which no single one is any longer predominant."[69] Clearly, the concept of labor-as-such, without restrictions, can be applied to labor in all societies universally in retrospect. Thus, the particularly bourgeois category becomes simultaneously the general concept of labor.

Marx's general concept of labor is, thereby, a consequence of the bourgeois society in which he was consciously situated, not the result of metaphysical dogma or an ontological faith. It would be a mistake to attribute here to Marx what he criticizes as ideological in others: the hypostatization of particular or even general categories into ahistorical, necessarily eternal concepts unrelated to the movement of history. Precisely with the concept of labor, Marx makes his points:

> The example of labor shows strikingly how even the most abstract categories, despite their validity – precisely because of their abstractness – for all epochs, are nevertheless, in the specific character of this abstraction, themselves likewise a product of historic relations and possess their full validity only for and within those relations. Bourgeois society is the most developed and the most complex historic organization of production. The categories which express its relations, the comprehension of its structure, thereby also allow insights into the structure and the relations of production of all the vanished social formations out of whose ruins and elements it built itself up, whose partly still uncovered remnants are carried along within it, whose mere nuances have developed explicit significance within it, etc.[70]

Abstract categories are products of historic relations. Further, their contradictory forms result from social disharmony: use value *versus* exchange

---

[68] *Ibid.*, p. 104, S. 24.

[69] *Ibid.*, p. 104, S. 25.

[70] *Ibid.*, p. 105, S. 25.

value, private property *versus* social production, particular *versus* abstract labor, wages *versus* surplus values; the conceptual tensions reflect class conflicts. Both the actuality of abstract labor and its expression in the concept of labor-as-such are products of the development of the historic preconditions of commodity production, which requires the severing in reality as in consciousness of the ties of workers to their land, tools, etc. as their own property. The concept of labor-as-such, already used by Adam Smith, is the central category of Marx's system, underlying his key analyses of the labor form of value, the antagonisms between labor and capital, the appropriation of surplus labor and the fetishism of the activity and conditions of labor in the products of labor. Marx's path of abstraction from the population of capitalist societies to abstract labor follows the well-worn trail of historical developments. Similarly, Marx's abstractions of production, value, etc. were in no way arbitrary, biased or metaphysical. They represented the conceptual appropriation of bourgeois social relations and they knew themselves as such in the sense that Marx's methodology – his path of abstraction and concretion, the manifold and reflected unity of his theory and social practice, his approach to history, his formation of general and specific concepts – transformed its social and historical conditions into its content. Hence the anticipatory character of Marx's abstractions in the early manuscripts: the process of abstraction at work in social research is itself guided by social theory. Only in *Capital* is the systematic circle complete in which the theory of bourgeois society justifies the abstractions which present the theory while within that theory the analysis of the fetishism of commodities explains the necessity of elaborating such a theory in the face of social mystifications.

A further point in the preceding quotation concerns the insight gained into pre-capitalist societies through the theory of capitalist society, that is the applicability of specifically bourgeois concepts to the interpretation of previous social formations. This trans-historical validity of many categories which are historically-specific to the present era – in that they were not fully developed and central in previous societies and hence were not prominent in the theoretical consciousness of those societies – is due to the fact that the central features of capitalist production, such as commodities, exchange, labor and money, were present in society in various forms. It is, of course especially useful to view pre-capitalist societies in terms of bourgeois categories when trying to understand them as stages in the prehistory of capitalism. But even more generally, the specific categories of bourgeois society give insight into previous societies in their sublimated form as general categories. But the insight isn't the whole picture, and what we are calling sublimation does cause an essential transformation.

An economist might, for example, generalize and say that production always requires an instrument (at least the hand of the worker) and accumulated labor (at least the training of a skilled hand). Joining these with a loose notion of capital, he might argue: "Capital is, among other things, also an instrument of production, also past, objectified labor. Consequently capital is a general and external relation of nature: that is," as Marx is quick to point out, "provided one omits just the specific quality which alone makes *instrument of production* and *accumulated labor* into capital."[71]   The point is that although the specific bourgeois category forms the basis for a general concept, there are crucial differences that must not be ignored. Labor under capitalism may well be conspicuously general in producing without concern for the particulars of the job or of the worker, it may clearly be a process of transforming nature into a product in order to support itself, but it is also *wage*-labor which does not own the preconditions of its labor or the product and which creates *exchange*-value in return for *monetary* wages – these latter characteristics differentiating it from all previous forms of labor. The specific – and therefore also the general – categories presuppose the historic process and a study of this process can clarify what distinguishes the specific from the general, second nature from the natural. Political economists tend to ignore the historical dimension and therefore end up as apologists for a more efficient version of what already exists anyway. They present production as "encased in eternal natural laws independent of history, and at the same time *bourgeois* relations are clandestinely passed off as inviolable natural laws on which is founded society *in abstracto*."[72]

The study of history often clears up the ideological illusions, not so much because of the details of the alternative view it may present, but merely in the fact that it uncovers the assumptions and maneuvers of the ahistorical approach. Marx uses historical knowledge to counter the claim that private property, for instance, is natural or necessary. He grants that production is always appropriation of nature, an individual taking possession, and thus property is (tautologically) a condition of production. But, he counters, "It is quite ridiculous to make a leap from this to a specific form of property, e.g., private property. (Which further and equally presupposes an antithetical form, *non-property*.) History shows, rather, common property (e.g., in India, among the Slavs, the early Celts, etc.) to be the (more) original form, and in the shape of communal property it continues to play a significant role for a long time."[73] This argument and the historical analysis of private property are carried out in the section of the *Grundrisse* to which we next turn.

---

[71] *Ibid.*, p. 86, S. 7.

[72] *Ibid.*, p. 87, S. 8f.

[73] *Ibid.*, p. 87f, S. 9.

To recapitulate: There are characteristics which are common to all stages of production and which are established by the mind as general characteristics; the so-called *general preconditions* of all production, however, are "nothing more than these abstract moments with which no real historical stage of production can be grasped."[74]

Retrospective Interpretation of the History of Property Relations

Marx's chapter on the "Forms which Precede Capitalist Production"[75] is divided into two sections: I) an historical reconstruction of the *prerequisites for wage labor* and II) an historical reconstruction of the *prerequisites for capital.* Schematically, that is, the economic system known as capitalism is defined by the relationship of wage-labor (embodied in the proletariat) to capital (embodied in the capitalist class), all other sectors having a secondary importance defined by their relation to the two primary classes and eventually dissolving into these classes with the progressive development and self-purification of the capitalist system, as Marx had already concluded in the 1844 *Manuscripts.*

Whereas the bulk of the *Grundrisse* is concerned with the *conditions* of capitalism, it is here the *preconditions* or *presuppositions* which are of interest. Marx had already drawn this distinction a few pages earlier in the *Grundrisse* in a section on primitive accumulation, with the example of the existence of cities. While a city can, as a city, reproduce its population, it cannot possibly have itself produced the founding population; they must have come from somewhere else. "While, e.g. the flight of feudal serfs to the cities is one of the *historic* (pre-) conditions and presuppositions of urbanism, it is not a *condition*, not a moment of the reality of developed cities, but it belongs to their *past* presuppositions, to the prerequisites of their formative process, which are suspended in their existence."[76] A "condition" of a city is, then, part of its permanent structure which insures its continued existence, whereas

---

[74] *Ibid.*, p. 88, S. 10.

[75] Marx's chapter on "Forms which precede capitalist production" is in the *Grundrisse*, p. 471-514, S. 375-415. An alternative, less literal translation is available in Karl Marx, *Pre-capitalist Economic Formations* (New York: International Publishers, 1969).

[76] "Forms," *Grundrisse*, p. 459, S. 363.

a "presupposition" is something that was required for the original creation of the city. A presupposition of wage labor is accordingly defined as an historical factor or development prior to capitalism which allowed for or contributed to the transformation of a large portion of the population into wage-laborers.

Marx opens his chapter by naming two such preconditions of wage-labor: "free-labor" and the separation or this "freed" labor from the ownership of the means and materials of labor:

> If one of the presuppositions of wage labor, and one of the historic preconditions for capital, is free labor and the exchange of this free labor for money, in order to reproduce and to realize money, to consume the use value of labor not for individual consumption, but as use value for money – then another presupposition is the separation of free labor from the objective conditions of its realization, from the means of labor and the material for labor. Thus, above all, release of the worker from the soil as his natural workshop – hence dissolution of small, free landed property as well as of communal landownership resting on the oriental commune.[77]

Marx proceeds to trace through history the relationship of the laborer to the land, the instruments, the means of consumption prior to production, and the laborer's own body. He follows these four relationships through tribal ("Oriental"), classical (Greek and Roman city-states), and feudal ("Germanic") society, treating the transformations, weakening and gradual demise of these relationships as a necessary prerequisite of capitalism. Originally, the worker, as a tribal member, is related to the direct communal property of land as his own. The intermediary stages see various combinations of communal and petty land ownership, until finally the laborer must pay rent to a landowner for use of land. The laborer's relationship to his instruments – even in the feudal guild system still retained along hereditary or craft lines – is finally dissolved with the factory system. The patriarchal systems also faded away, which had cared for the members of the tribe, extended family, estate or guild during the process of production. Finally, the system of slavery and serfhood came to an end, leaving *free-labor* separated from the means of production and waiting to be appropriated through the exchange of wages for labor time.

In understanding the economic formations which precede capitalism as moments in the socio-historical process which provided the preconditions of capitalist production, Marx is far from inventing a mythic justification for the

[77] *Ibid.*, p. 471, S. 375.

present system by deducing it as the final stage in a conceptual development; on the other hand, he is not projecting the character of the present system back into pre-history. Rather, his concerns are based in the present, but he strives to show how the past was essentially different and then to outline the logic of the actual historical development which made capitalism possible and actual, it is a *reconstruction* in the sense that it starts out from the end of history (so far), seeking to understand that end as the result of an historical process and to understand history as leading up to the present.[78] The resultant view of the present in terms of its formative process (as historically mediated) stands critically opposed to the (ideological) view of the given as it immediately appears.

Marx's historical analysis provides the handle for critically grasping the bourgeois ideology of "just exchange," the view that wages which represent accumulated labor-time are exchanged for an equivalent amount of the worker's living labor-time. Viewed from within the capitalist system, the presuppositions of the theory of just exchange may well seem obviously true; but considered in terms of the historical prerequisites of the entire system, new presuppositions are uncovered and the theory is shown to be false. Private property is not the result of the capitalist's own labor, but is rather the result of removing the conditions of labor (land and tools) from the control of the worker. In this context, the theory of just exchange is seen to be ideology: socially necessary illusion which obscures the process of unjust appropriation.[79]

Bourgeois social theory ahistorically projects capitalist relations back to a "state of nature" in which, ironically, everyone belonged to the proletariat, i.e. embodied free-labor deprived of property in the conditions of production. Marx counters with a diametrically opposed reading of history, which he documents. His argument, which we will now try to follow, runs roughly as follows: At the dawn of human social history, as nomadic life came to an end and social formations evolved, agricultural production was carried out communally. Each member of the community shared in the ownership of the

---

[78] The following comment opens the seventh of Benjamin's "Theses on the philosophy of history," which goes on to show how a non-Marxian, historicist attitude toward the past inevitably sides with the present ruling class. Cf. Walter Benjamin, *Illuminations* (New York: Schocken, 1969), p. 256:

> To historians who wish to relive an era, Fustel de Coulanges recommends that they blot out everything they know about the later course of history. There is no better way of characterizing the method with which historical materialism has broken.

[79] Cf. *Grundrisse*, p. 509, S. 409.

means of production, the communal land and the shared tools. Through the historical development which eventually culminated in capitalism, the relationship of the worker to the land he worked took on many forms, leading generally to a dissolution of the organic unity of the worker and owner. The conditions of production were gradually accumulated into the hands of a few (the capitalists), leaving the masses property-less and dependent upon others for work. The formation of the individual – a laborer stripped of defining ties to land and community, a numerical unity on the competitive job market – was an historical result and a precondition of capitalism, not a natural, a-temporal phenomenon *a la* Robinson Crusoe. The same process which transformed people in social settings into abstract, inter-changeable individuals made the products of labor into abstract, arbitrarily exchangeable value, commodities. The relations of capitalist production, commodity relations, pervaded every aspect of social existence, through their two-fold character obscuring the nature of social reality.

The formation of a capitalist class, on the other hand, is likewise not the result of a natural, ahistorical merely quantitative inequality – that due to physiological or geographical differences some people have more wealth than others. Wealth can only count as capital in a social setting in which some people are property-less and others own sufficient property to provide others with materials of labor. The separation of workers from property as well as the development of industrial production skills are results of the historical process which precedes capitalism and their evolution is not to be explained in terms of capitalist relations as the theory of just exchange would have it. Marx is especially concerned with the transition from feudalism to capitalism because it is here that the capitalist first appears – taking advantage of the circumstances, rather than supporting the whole. The analysis is succinctly outlined by Marx:

> What enables monetary wealth to turn into capital is on the one hand, that it finds free laborers, and on the other, it finds means of subsistence, materials, etc., which previously were in one form or another the property of the now objectless masses, and are also *free* and purchasable. However, the other condition of labor – a certain level of skill, the existence of the instrument as means of labor, etc. – is already available in this preparatory or first period of capital. This is partly the result of the urban guild system, partly of domestic industry, of such industry as exists as an accessory to agriculture. The historic process is not the product of capital, but its presupposition. By means of this process the capitalist then inserts himself as (historic) middleman between landed property, or property generally,

and labor. History knows nothing of the congenial fantasies about capitalist and laborer forming an association etc.; nor is there a trace of them in the conceptual development of capital.[80]

Contract theories of social institutions perpetuate illusions about history and the place of capitalism in history. They explain communal, social and economic relations as decisions of rationally calculating individuals. For Marx, it is precisely rational calculation and the individual which must be accounted for as historical products of a process which began with naturally, spontaneously existing human communities and their attendant social institutions and economic arrangements. Surveying history with an eye to the uniqueness of capitalist relations, Marx finds two major phenomena common to the various pre-capitalist, pre-industrial forms of society: *property* in the conditions of one's labor and existence as a member of a *community*. Having traced the evolution of society from the nomadic tribe through oriental despotism and classical city-states to the feudal Middle Ages, Marx stresses that:

> The main point here is this: In all these forms – in which landed property and agriculture form the basis of the economic order, and where the economic aim is hence the production of use values, i.e. the *reproduction of the individual* within the specific relation of the community in which he is its basis – there is to be found: (1) Appropriation not through labor, but presupposed by labor; appropriation of the natural conditions of labor, of the *earth* as the original instrument of labor as well as its workshop and repository of raw materials. The individual relates simply to the objective conditions of labor as being his; . . . (2) but this *relation* to land and soil, to the earth, as the property of the laboring individual – who thus appears from the outset not merely as laboring individual, in this abstraction, but who has an *objective mode of existence* in his ownership of the land, an existence *presupposed* by his activity and not merely as a result of it, a presupposition of his activity just like his skin, his sense organs, which of course he also reproduces and develops, etc. in the life process, but which are nevertheless presuppositions of this process of his reproduction – is instantly mediated by the naturally arisen, spontaneous, more or less historically developed and modified presence of

---

[80] *Ibid.*, p. 505, S. 404f.

the individual as *member of a community* – his naturally arisen presence as member of a tribe, etc.[81]

This passage presents the basis for Marx's analysis in his chapter, contrasting all pre-capitalist economic formations as a group to capitalist social relations. By viewing the transition from feudalism to capitalism – prepared for by all previous social history – as the great historical watershed, Marx stresses the uniqueness of capitalist relations, their socio-historical specificity, their unnaturalness.

Many important themes of Marx's thought are related to this passage. Property is here defined in terms of the means of production and private property is seen as derivative of communal property. The bourgeois conception of the individual in abstraction from his social setting is taken as a result of social and intellectual history; its claim to *a priori* status is rejected. People have an objective existence in their community – as in their more or less historically developed language and body – which is the precondition of their labor, not first a result of objectification. The whole analysis reflects Marx's radically historical approach as well as his driving concern with understanding capitalism as a moment in the flow of actual history. Accordingly, both the evolution and the reproduction of society are understood as integral facets of social production. This historical approach suggests a new formulation of the theory of alienation, one which clearly contrasts alienation in capitalist commodity production with the organic ties of the worker in pre-capitalist economies to the means and materials of his work, to the product as an immediate use value and to himself and his community. These important themes must now be developed.

History has teleological meaning – but only for us from our retro-perspective, not mystically in itself. The production of history has, according to Marx, been largely a byproduct of the reproduction of given situations through the production of use-value. Given any agricultural society, the goal of production in that society was self-perpetuation of the community through reproduction of the race and of everyday life. But the natural attempt to maintain the status quo, particularly in the more advanced social forms, can itself force change; the fates of wars of defense have resulted in the rise and fall of untold numbers of nations. One of Marx's favorite examples is the timely one of population expansion. It is a common observation in anthropology that certain tribal structures can perpetuate themselves indefinitely and self-sufficiently given a constant population and no external interference. But suppose that the attempt to reproduce the population becomes too successful. Then the social relations, division of labor or form

---

[81] *Ibid.*, p. 485, S. 384f.

of production would be forced to change or die. The details of a society's atemporal structure can thereby become central to the dialectic of its historical development.

> For instance, where each individual is supposed to possess so many acres of land, the mere increase in population constitutes an obstacle. If this is to be overcome, colonization will develop and this necessitates wars of conquest. This leads to slavery, etc. Also, e.g., the enlargement of the *ager publicus*, and hence to the rise of the Patricians, who represent the community, etc. Thus the preservation of the old community implies the destruction of the conditions upon which it rests, and turns into its opposite.[82]

The conservative impulse can remain at the base of truly revolutionary developments – at least within agricultural society.

When the relations of production change in accordance with a modified mode of production, there is not merely a change in relations between unchanged people; people's very natures are transformed. "The act of reproduction itself changes not only the objective conditions – e.g. transforming village into town, the wilderness into agricultural clearings, etc. – but the producers change with it, by the emergence of new qualities, by transforming and developing themselves in production, forming new powers and new conceptions. new modes of intercourse, new needs, and new language."[83]

Language is an intriguing topic for Marx because of its totally historical and social character. Both a presupposition and a condition of history, language is in turn modified by new social development. A person's language is both a prerequisite and an expression of his membership in a community: "As regards the individual, it is clear, e.g., that he relates even to language itself as *his own* only as the natural member of a human community. Language as the product of an individual is an absurdity. But so also is property."[84] The point of Marx's references to language is that *property* has the same characteristics which clearly belong to language. An isolated individual could farm a plot of land and *possess* tools – assuming he had been trained in farming and in the construction and use of tools – but we would not call the land and tools his *property*.

---

[82] *Ibid.*, p. 493f, S. 393.

[83] *Ibid.*, p. 494, S. 394.

[84] *Ibid.*, p. 490, S. 390.

Property is a social (legal, interpersonal) relationship defined by the exclusion of other persons from certain rights. Originally, proprietary claims took place between tribes, neighboring communities who claimed areas of land or natural resources as their property to the exclusion of other tribes. The tribal members shared the fruits of such property, which they owned as members of the community. They may have possessed homes, clothing and such individually, but the land, tools, weapons, aqueducts, seeds and technical knowledge were communal property. In later social systems, such as the Roman Empire, communal property was privately possessed – to be a Roman citizen and to possess Roman land were synonymous – or some combination of private and public lands was established.

Property refers, as the word etymologically suggests, not only to what one (legally) owns, but to what is (essentially) one's own, what is "proper" to oneself. Throughout precapitalist history, people were defined physically and mentally, objectively and subjectively, economically and socially by their membership in a community. Marx thus concludes, *"Property* therefore means *belonging to a clan* (community) (having one's subjective/objective existence in it); and by means of the relation of this community to the land and soil, (relating) to the earth as the individual's body, there occurs the relationship of the individual to the land and soil – for the earth is at the same time raw material, tool and fruit – as to a presupposition belonging to his individuality, as its modes of his presence. We *reduce this property to the relation to the conditions of production."*[85]

If people are socially defined, then society in turn can be viewed in terms of the social relations which are primarily structured according to the mode of production of the producing subjects. The primary relationship is that of work: productive people relating to nature. Work is socially mediated – people produce as members of communities and the structure of the communities is reciprocally conditioned by the mode of production. Property, too, is an expression of the work relationship, always defined historically in relation to the prevailing mode of production. The first half of Marx's historical considerations accordingly conclude that *"property* – in its Asiatic, Slavonic, ancient classical and Germanic forms – therefore originally signifies a relation of the working (producing) or self-reproducing subject to the conditions of his production or reproduction as his own. Hence, according to the conditions of this production, property will take different forms."[86]

---

[85] *Ibid.,* p. 492, S. 392.

[86] *Ibid.,* p. 495, S. 395.

The result that Marx arrived at through analysis of capitalist relations in the *Manuscripts* – that property is historically-specifically defined, expressing the prevailing mode of production – is here reached through a study of pre-capitalist history. The new approach sheds fresh light. Not only is the illusory atemporal aura surrounding bourgeois private property discarded in showing the relation of *private* property to *commodity* production in particular, but in drawing this relationship, the unique character of private property is revealed. Bourgeois private property differs essentially from all pre-capitalist forms of property, which were variations on communal property. "Private" property is a privative form in that the worker no longer relates through the property institution to the conditions of his labor as his own. The opposition of labor to capital means precisely that the worker is not the owner of the objective basis of his existence – and conversely the owner of the conditions and product of labor is not the worker who transforms the one into the other. The split of the original worker/owner unity is a fundamental expression of the alienating character of capitalist relations.

Dealing with alienation in terms of the worker/owner division of labor underlying capitalism has advantages over the view in terms of the worker's loss of his product, although the two approaches are in the end equivalent. It may, for instance, be less tempting to psychologize the worker's frustration at self-objectification gone sour and keep the objective societal configuration in the fore.

Of primary significance is the clarity with which the historical character of alienation appears. Alienation as caused by a dichotomy between worker and owner is seen to be "unnatural," i.e. unique to the bourgeois era and foreign to the simpler social forms. Accordingly, one can even recognize the relation of alienation to the destruction of the conservative values: sense of community, rootedness in the land, treasuring of tradition, pride in work-manship, etc. A closer study of alienation as an historical development, distinguishing losses which must be recouped from those which represent valuable progress and potential may suggest goals for a future society – one chosen more consciously and democratically than past social systems – and potentials for realizing them.

How then did capitalist alienation come about? Whence the split in the previous unity of worker and owner? If it is not true that there have always been some people with property in the means of production and other, property-less people who are therefore dependent on the first group for jobs, then how did the capitalists come to acquire a monopoly on property to the exclusion of the workers? The ideology would have it that the capitalists themselves created their property through diligent labor and accumulated their wealth through careful saving – in contrast to the allegedly lazy and wasteful masses. Marx vigorously rejects this view and argues that the so-called primitive accumulation of capital was actually a *rearrangement* of existing property. The redistribution and centralization of property – of the means, materials and conditions of production – was made possible by developments in the mode of production: urban craft production in the guilds, an international money system in merchant trade, technical changes in spinning and weaving, etc. Property, or wealth, which became the capitalist's in the transition from feudalism to capitalism, had already existed and had merely become flexible enough to be redistributed through all manner of usury, trade, hoarding, politics, trickery and force.

> Nothing can therefore be more ridiculous than to conceive the *original formation* of capital as if capital has stockpiled and created the *objective conditions of production* – food, raw materials, instruments – and then offered them to the *dispossessed* worker. What happened was rather that monetary wealth partly helped to dispossess the labor powers of the individuals capable of work from those conditions; and in part this process of divorce proceeded without it. . . . As to the *formation of monetary wealth* itself, before its transformation into capital: this belongs to the prehistory of the bourgeois economy.[87]

Spinning and weaving, Marx's clearest illustration, was one of the first jobs to be transformed from the home into the factory. Traditionally, each family unit had its own spinning wheel and loom with which the family met its own clothing needs, as one of its many productive activities. After the transformation, certain individuals produced cloth to be sold as their sole productive activity, using the wages earned thereby to meet their own specific needs. Capital had neither invented nor manufactured the spinning wheel or loom; clothing had previously been produced and consumed. Capital neither created nor stockpiled the necessities of life which now had to be purchased. "The only change was," Marx points out, "that these necessities were now thrown onto the *exchange market* – were separated from their direct connection with the mouths of the retainers etc. and transformed from use value into exchange values."[88]

The change from a worker/owner unity to the distinction of property-less workers and non-productive owners meant a shift from concrete, immediate use-values to abstract, socially mediated exchange values. On the one hand, this shift away from use-values as such is a prerequisite of capitalist

---

[87] *Ibid.*, p. 509, S. 408.

[88] *Ibid.*, p. 507, S. 407.

accumulation and deprivation: only so much use-value can be accumulated and still be meaningful, while no one can be deprived totally of use-values and continue to live. On the other hand, the shift is also a result of capitalism, or is reinforced by the spread of capitalist relations: someone who wants to buy commodities must earn money and one who earns money must meet his needs through the exchange of commodities. The totalizing character of capitalism is thus built into the nature of commodities, i.e., it is part of the dynamic of the social relations of commodity production.[89] The so-called primitive accumulation of capital is to be accounted for in terms of commodity production and its demands. Commodity production, based on monetary quantity, is qualitatively "free." The conditions of production become a "free fund," liberated from their ties to particular individuals or families. The individuals thereby deprived of their property simultaneously become potentially "free wage-laborers," obliged to sell their labor. For Marx, "This much is evident: the same process which divorced a mass of individuals from their previous affirmative relations to the *objective conditions of labor*, negated these relations and thereby transformed these individuals into *free laborers*, is also the same process which freed these *objective conditions of labor* – land and soil, new material, necessaries of life, instruments of labor, money or all of these – potentially from their *previous state of attachment* to the individuals now separated from them." Further, "Closer inspection will show that all these processes of dissolution mean the dissolution of production in which use-value predominates production for immediate use." And finally, "Again, closer examination will also reveal that all the resolved relations were possible only with a definite degree of development of the material (and therefore also of the intellectual) productive forces."[90] The four historical tendencies go hand in hand: (1) from owning-worker to free-labor or wage-laborer, (2) from worker's property to private property, (3) from production of use-value to production of exchange value or commodity production, and (4) from agriculture to manufacture. Once these processes, which Marx sees as stretching in various patterns across the entire span of history, have completed their transformations, the scene is set for capitalist production, which has, indeed, already set in as part of the historical process.

The conjuncture of these four processes in Marx's account indicate how, as already suggested above at the start of Chapter V, his reconstruction of the history of property relations, understood in their relation to the prevailing - modes of production, provides a unity to Marx's doctrines, presents the core of his historical materialism and supplies the critical fulcrum for his critique of political economy. The thematic unity underlying the altering styles or ap-

---

[89] Cf. *ibid.*, p. 511, S. 410f.

[90] *Ibid.*, p. 502f, S. 402.

proaches of Marx's various writings can be summed up in the term "commodity production," a concept which is central to all his analyses of capitalism, whatever vocabulary or approach he may be using. Historical materialism means, for Marx, a primary concern for social mediations, a radically historical methodology and an eye on the prevailing mode of production as the primary determinant of an historical epoch. Arguing against non-historical theories of just exchange, the eternal nature of man and the productivity of the capitalist, Marx puts political economy on a philosophically critical and politically radical basis.

The *raison d'etre* of Marx's historical approach and of his critical social theory in general makes a sudden appearance in the midst of his historical account – in a passage charged with excitement and pathos. The emotions are appropriate to the paradox that defines capitalism's status in history: undreamed of potential repressed. The freedom implied by the phrase "free labor" was not wholly sarcastic. The feudal peasant is not freer than the wage-laborer even though he is not alienated and exploited in the same way. The peasant is chained to his plot of land and narrow occupation by the primitive mode of production as well as by corresponding sentiments. In comparison, the industrial worker has a universal potential: he could produce in practically any manner imaginable, thanks to the developments in production resulting from capitalism – if, that is, capitalism did not at the same time confine him within a physically and intellectually crippling division of labor and system of alienation. Capitalism, which made man the universal creator, reduced people to one-sided drudges.

Marx rejects the short-sighted criticism of capitalism which argues that now people work for money or live to work where they once worked for goods, for life. What is money? Marx asks, What is abstract value? – when the limited bourgeois form is stripped away – if not "the universality of individual needs, capacities, pleasures, productive forces, etc. created through universal exchange? The full development of human mastery over the forces of nature, those of his own nature as well as those of so-called nature?"[91] In bourgeois political economics – and in the epoch of production to which it corresponds – this complete elaboration of what lies within man appears as a complete emptying-out, this universal objectification as total alienation, and the tearing down of all limited, one-sided aims as sacrifice of the (human) end-in-itself to an entirely external end. The origin of Marx's definition of man as *homo faber universalis* in the *Manuscripts* – and of the anti-capitalist desires to achieve human creative potential today – can be traced to the economics of the era of capitalist production, which is immanently criticized for suppressing the possibilities it has developed. The introduction of abstract value and abstract

[91] *Ibid.*, p. 488, S. 387.

("free") relations in general on a pervasive scale has abolished provincial limitations, creating "social value" which joins all aspects of life into a worldwide social totality. At the same time, however, it subordinates use-value to calculations of private exchange value (profit), rather than raising them to the level of "*social use*-values." The alienation of historically produced human nature and the repression of attempts at human liberation are to be understood in terms of the two-faced progress represented by commodity relations. Far from basing itself on theological or Victorian faith in the necessary progress of humanity, Marx's analysis incorporates an insight into the self-annihilation of progress to date. To trace this stunted dialectic of value in hopes of freeing it from the fetishism of commodities is the task of *Capital*.

Chapter VI. Presentation: *Capital*

The dialectic of enlightenment has reached a contradictory stage in the era of capitalism, an economic system which creates the potential for a new humanity while exploiting people and which universally propagates an un-dreamed of wealth of knowledge in the form of mass deception.[92] *Capital*, an analysis of the system of commodity production in its ideal conceptual purity, provides clarifications which lead to a comprehension and calculus of its emancipatory and repressive tendencies. The two-faced rationalizations of bourgeois political economy – ideology in an emphatic sense – are traced by Marx to the bifurcated nature of commodities and of their production. Conceptual, social and historical analysis dispels the confusions which characterize the fetishism of commodities and paints the way to realizing the un-kept promises of our epoch. The capitalists' most authentic language, the equations of profit, is transformed into indices of exploitation and calculations of potential human progress by questioning the foundations of capital which had remained presupposed but unspoken.

---

[92] Cf. Max Horkheimer & Theodor W. Adorno, *Dialectic of Enlightenment* (New York: Herder & Herder, 1972); *Dialektik der Aufklärung* (Amsterdam: Querido, 1947).

The Form of Value of Commodities

A *repeated* conclusion of the preceding interpretation of Marx's work has been that the realm of production has a priority in the analysis of society and that an analysis of capitalism must therefore begin with a consideration of commodity production. Part I of the first volume of *Capital* is accordingly on "Commodities and Money." The discussion takes place in terms of a theory of commodity value, designed to clear the confusions surrounding the nature of money.

In his original preface to *Capital*, where he warns of the difficulty of reading Part I, Marx likens his economic science to biology and chemistry, which study the elementary cell or molecule in order to understand large complex bodies. In Marxian economics, however, the "force of abstraction" must replace the tools of microscope and chemical reagents, increasing the conceptual strain. Continuing the analogy, Marx states that in capitalist society "the commodity form of the product of labor or the form of value of the commodity is the economic cell form."[93] Thus, the starting point for a systematic presentation of capitalist production is the commodity form of products and the form of value of the commodity.

Several models of systematic presentation useful in following Marx's arguments suggest themselves. The extreme contemporary form of systematization is axiomatization, in which one begins with formal definitions and mathematized axioms, proceeds through a hierarchy of theorems derived from the axioms and finally provides an interpretive scheme relating the most derived propositions to reality. Marx's analysis of the way in which the quantitative value of a commodity varies under different conditions is easily adapted to this mathematical approach.[94] A parallel technique is the dialectical method Hegel employs in his *Science of Logic* which, for instance, analyses the mutual determinations of what he calls "reflex categories." This approach is very much at work in Marx's development of the *form* of value. However, as opposed to mathematics and idealism, Marx's work is materialistic and political. Materialistically, the presentation of commodity production stresses the historical specificity of this mode of production and grounds the analytic categories in their historical object. As political, *Capital* strives to reveal the situation of people in society – a task necessitated by the tendency of commodity relations to obscure social, interpersonal relations.

---

[93] Karl Marx, Capital, p. 8, S. 1f.

[94] Cf. *ibid.*, p. 53f, S. 34f.

Capital's path from the simple abstraction of the commodity to the complex concreteness of capitalist production as a totality is at every level teleologically determined by its real object, as this had been encountered in the concrete history of class struggles and in a succession of pre-Marxian theories of surplus value. Of particular interest in relation to continental philosophy from Hegel to Heidegger, is the way in which Chapter 1, which culminates in the revelation of the secret of commodity fetishism, is motivated by the misleading character of the form in which the value of commodities "appear."

Marx's discussion of the form of value, his most original theoretical contribution to the analysis of capitalist economics, is the most difficult section of *Capital*. The history of Marx interpretation can be viewed as a kaleidoscope of misinterpretations of this section. Marx was well aware of the problem and posted warnings repeatedly. On the first page of text in *Capital*, in his Preface to the first edition, preceded only by a note on the circumstances of publication, Marx warned:

> All beginnings are difficult; this goes for every science. The comprehension of the *first chapter*, namely the section containing the *analysis of commodities*, will therefore present the most difficulties. That which specifically concerns the *analysis of the substance of value* and of the *magnitude of value*, I have popularized as much as possible. Not so with the analysis of the *form of value*. This is difficult to understand because the dialectic is much sharper than in the first presentation. I therefore advise the reader who is not thoroughly accustomed to dialectical thinking to completely skip the section from p.15 to p.34 and instead to read the appendix at the end of the book, "The Form of Value."

> There I have tried to present the matter as simply and even
> as didactically as its scientific nature allows. After finishing
> the Appendix, the reader can proceed with the text at p.35.
> The *form of value*, whose completed *Gestalt* is the *money form*, is
> without content and simple. Nevertheless, the human spirit
> has sought in vain for more than 2000 years to get to the
> bottom of it.[95]

In later editions and in the English translation, the section of sharp dialectics and the reference to it in the Preface have been deleted and the appendix has been moved into the text as the bulk of section 3. The popularized replacement and its translation are so commonsensical that it is easy to overlook what remains of the "sharp dialectic" and to misjudge this cornerstone of Marx's theory.

Especially if one is to consider *Capital* as a philosophical statement adhering to rigorous methodology, it is imperative to follow the first edition's analysis of the form of value. For it is here that Marx unfolds his dialectical materialist presentation of abstract labor, as the defining characteristic of capitalist production. Once comprehended, this presentation preserves the revolutionary content of the labor theory of value from bourgeois revision, not least of all by revealing the secret of commodity fetishism. The genetic demystifying de-construction of fetishism disarms the power of commodity relations to obscure their origin in the labor of the working class. Accordingly, the following discussion of *Capital* will limit itself to retracing Marx's sharp dialectic from the definition of the commodity, through the establishment of abstract labor, to the domination of fetishism.

Through an understanding of the significance of this dialectic, the separation of Marx's work into an interpretive (qualitative) outlook and an explanatory (quantitative) calculus can be made more precise in terms of the analysis of the substance, magnitude and form of value. The view that human labor is the *substance* of all value provides a basis for proletarian "ideology." Economics which calculates the *magnitude* of value on the basis of the quantity of socially necessary labor time embodied in a product can serve as a corresponding "science." Marx's approach is not strictly opposed to these, but neither is it limited to them. Marx's analysis of the *form* of value – which presupposes neither its substance nor its magnitude – provides a basis for what would otherwise be "mere ideology," a perspective chosen on the basis of self-interest alone. By comprehending the social constitution of abstract labor as an historically developed category, the analysis of the form of value

---

[95] Karl Marx, *Das Kapital*, Bd. I (Hamburg: Meissner, 1867), S. viif. Cf. *ibid.*, p. 7f, S. 1.

provides the theoretical precondition for defining the substance and magnitude of value in terms of abstract human labor.

An ideology of labor makes the same false move as the ideology of capital: absolutizing categories of the present era as eternal. Marx's goal is not the worship of labor – it was not he who coined the concentration camp slogan, "work makes one free" – but the just reward for the unfortunate necessity (un-freedom) of work. Just reward according to the theory in *Capital* consists in the worker owning the surplus value produced by his work or, less simplistically, a society of producers jointly owning the full product of their social labor. The theory of surplus value which spells out this goal takes labor as the substance of value and labor time as the measure of the magnitude of value. But it is the analysis of the form of value as historically specific which lends the critical sharpness to the socialist goal as anti-capitalist, as historically situated.

The common impression is thus wrong, that Marx's primary theoretical contribution is the proposition that commodity prices are determined by the labor time required for their production. Even this proposition does not, as Heidegger has suggested, claim that labor is the substance of commodities, but rather that *abstract* labor is the substance of the *value* of commodities. However, even the correctly understood economic proposition is not peculiar to Marx. John Locke and William Petty had long before argued for such a labor theory of value and Adam Smith and David Ricardo had also accepted it. Only when the capitalist class had established itself over the aristocracy, did economists fear that the labor theory could be transformed from a bourgeois to a proletarian ideology, a possibility forcefully symbolized by the Paris Commune of 1848. Marx then championed the labor theory, but he did not present it with much fanfare in *Capital*.

In its simple form, the proposition that labor is the substance of value has little need of argumentation. Strictly speaking, this proposition is not only non-deductive, but for Marx it is neither descriptive nor prescriptive. It neither describes the appearance of the prevalent pricing scheme, nor simply provides a measure of values for a future society. As a theoretical abstraction, labor value grants social theory a tool for comprehending and distinguishing various aspects and categories of value. The critic of Marx's theory must, therefore, answer to the power of Marx's system to analyze capitalist society as well as respond to the arguments in Marx's earlier writings and in Locke and others. It must of course be remembered that Marx never says actual prices are simply equivalent to a measure of labor time. Their "determination" (in the non-mathematical sense) by labor time is merely the first of many conditioning factors – first not so much in quantitative importance as in the chain of increasingly complex explanatory analyses presented in *Capital*.

Marx's approach to the analysis of value is scarcely oversimplified by starting with a statement from near the start of Chapter I of *Capital*. "We know now," says Marx already, "the *substance* of value. It is labor. We know its *magnitude*. It is *labor time*. Its *form*, which stamps *value* as *exchange* value, remains to be analyzed."[96] This task of analyzing the form of value is indeed what remained for Marx to contribute to political economy. In a footnote concluding his presentation of this analysis (retained in a different context in later editions), Marx outlines the reasons for the previous neglect of this task as well as for its critical import:

> It is one of the fundamental failings of classical political economy that it never succeeded in discovering, on the basis of analysis of the commodity value, the *form* of value, which makes it into *exchange* value. Even in its best representatives, such as Smith and Ricardo, economics treats the form of value as something completely uninteresting or external to the nature of the commodity itself. The reason is not simply that their attention is completely absorbed by the analysis of the magnitude of value. It lies deeper. The *form of value of the product of labor* is the most abstract, but also most general *form* of the *bourgeois* mode of production, which is thereby characterized as a specific kind of social mode of production and is thus simultaneously historically characterized. If one therefore mistakes it for the eternal natural form of social production, than one necessarily also misses what is specific in the form of value, that is, the commodity form, further developed as the money form, the capital form, and so on. One can thus find among economists who are in thorough agreement about the determination of the magnitude of value by labor time the most varied and contradictory conceptions of money.[97]

Before proceeding with his analysis of the form of value, Marx clarifies the distinction between use value and exchange value. The use value of a commodity is its utility in satisfying human needs. It is a function of the physical properties of the commodity considered as a natural object. Although use value does depend upon the social development of human needs, it is a basically trans-historical category. Exchange value, on the other

---

[96] *Ibid.*, S. 6.

[97] *Ibid.*, S. 34f.

hand, is the value of a commodity in exchange, on the market. It is a thoroughly social category, specific to exchange economies. Commodity production is defined by the congruence of use value and exchange value in the product. Production for the immediate satisfaction of one's own needs is not considered "commodity production." To produce commodities one must produce use values for others, social use value, products whose utility is only realized through the medium of exchange.

This kind of definition, distinction and clarification could be made by any Anglo-American philosopher and is not a primary accomplishment of Marx. However, in addition to stressing the *social* character of exchange value against the confused ideologues who saw the exchange value of a commodity as a natural property, Marx applied the distinction to *labor*. This latter point, that the labor embodied in a commodity has the same two-fold character as the commodity produced is, according to Marx, the fulcrum around which the comprehension of political economy revolves. Marx was the first to develop this point critically. The theory of surplus value developed in *Capital*, which explains the exploitation of labor by capital, is based on this point. That is, the contract between capitalist and laborer, which claims to exchange equivalents, actually pays wages equivalent to the labor power's exchange value (subsistence) in exchange for its considerably greater use value (productivity). The revelation of this injustice in the exchange of wages for labor power is, indeed, the fulcrum for Marx's critique of political economy.

According to Marx's theory of surplus value, then, the social antagonism expressed in the class conflict between labor and capital is based on the distinction between the use value of labor and its exchange value. The social antagonism is thus fundamental to bourgeois production, in which both labor and its products are determined in opposing ways. For, there are not two separate kinds of labor embodied in the commodity, but rather, the labor is variously and even incompatibly "determined" (*entgegengesetzt bestimmt*). This is the contradiction between use value and exchange value, which is the fundamental contradiction of capitalist society and which appears in the form of many social contradictions, antagonisms and crises.

The potentially misleading term, "contradiction," is to be understood in the sense that production is subject to criteria which make incompatible demands. In necessarily compromising on each criterion under the pressure of competing criteria, the desired effect is altered. The contradiction develops under its own impetus in directions not simply related to any one of the original criteria. For instance, the drive to produce exchange value may either eliminate (as unprofitable) the production of certain socially urgent use values or it may encourage over-production beyond the limits of consumption (as use value), resulting eventually in economic depression. Alternatively, technological progress, which is encouraged by the drive for private profit, exerts a social tendency to lower the rate of profit in a self-defeating interplay of private and social determinants.

Another way of looking at the capitalist contradiction is thus in terms of the antagonism of *private* and *social* constraints on commodity production. In pre-capitalist, non-commodity production, the individuals (or family or tribe) themselves produced what was directly necessary for themselves. Production was carried out privately in accordance with private criteria or else it was carried out communally in accordance with purely communal requirements. In commodity production, production is organized privately by the capitalist who hires wage laborers to supply the labor power and it is organized for the private end of accumulating capital. However, the means of achieving the end have a social character and the private production must fulfill social conditions. The commodity must be sold on the market in order for its exchange value to be realized: it must thus be a social use value; it must also be competitive with other commodities on the market in terms of quality and price; this means the efficiency of the productive techniques must be up to the social norm; and so on. The dynamic engendered by the definition of commodity production as production for private gain through social exchange can be seen in the macroeconomic development of capitalism out of feudalism and into crisis as well as in the microeconomic relations of the steady-state system.

The historical dynamic also reveals the material pre-conditions for the harmonious economic system implicit in a critique of the contradiction between use value and exchange value, private production and social exchange. A post-capitalist society would, that is, have to remove the antagonism of exchange to human utility by socializing the processes of production and consumption through a non-capitalist and non-private distribution of both the means of production and the results of production. At least some of the preconditions of such a social transformation are already given as results of the development of capitalism: a world market, enormous productivity, advanced communications, computerized information processing, centralized economics and, last but not least, the threats of the consequences of continuing to exist under capitalist relations and crises: nuclear war, starvation in the third world, irreparable ecological damage, fascist governments and in general the repression of free human activity as it is now possible.

The cornerstone of Marx's analysis of the capitalist contradiction is the sharp dialectic of the form of value of the commodity form of the product of labor. The analysis of the form of value follows a dialectical format because it is an analysis of the mutual mediations of moments in the relation which defines the form of value of commodities. The analysis parallels Hegel's famous master/slave dialectic in his *Phenomenology of Spirit* because both the relation of the form of value and the relation of master to slave are, like the relation of left to right, examples of what Hegel calls "reflex categories." You cannot have the one without the other.

The characteristic of commodities, versus products of pre-capitalist production for immediate consumption, is that they are exchanged. Furthermore, they are exchanged in specific quantitative proportions. This emphasis on quantity is characteristic of capitalism, distinguishing its systematic exchange from the more sporadic, qualitative and variable trade and barter which plays a secondary role in pre-capitalist economics. The simple, first or relative form of value of the commodity is thus:

$$x \text{ of commodity } A = y \text{ of commodity } B$$

or, e.g., 20 yards of linen = one coat

or twenty yards of linen are worth one coat.

The form of value is the equation. The equation posits quantities of different kinds of commodities as equivalent. The value of the one commodity is reflected in the value of another and neither has a value outside of such equations or relations of exchange. The common denominator of the commodities (their substance) which allows them to be equated in definite proportions is labor. In equating different types of products of labor, the form of value equates the different types of labor expended. The form of value is thereby the source of abstract human labor as such. The next section will follow this analysis of the form of value as presented in the first edition of *Capital*.

Abstract Labor in Theory and Practice

"The *real* relation of commodities to one another is the *process of exchange*."[98] The relative form of value given by the equation, x of A = y of B, is the *abstract* representation of the concrete historical, social process of exchange which takes place in terms of money; this elemental form of value is "in a certain sense the cell form or, as Hegel would say, *the in-itself of money*."[99]

---

[98] *Ibid.*, S. 44.

[99] *Ibid.*, S. 15.

In the language of the *Grundrisse*, the relative form of value is the ultimate abstraction, from which a systematic presentation of the concrete totality can begin. Later in *Capital*, Marx deals with the real process of exchange, but in Section 1 he is working on a high level of abstraction comparable to that of Hegel's *Logic*. Unlike Hegel, however, Marx conceives of the abstract determination as itself a *result* of the concrete processes which it analyses. The abstract is only a "source" of the concrete in theoretical presentations, not in real historical processes. Recognizing this, Marx was consistently able to understand Hegel's system better than Hegel had, namely as an analysis of the logic of the capitalist world as opposed to a description of actual developments. Marx's Hegel critique is as important to his analysis of abstract labor as his critique of Feuerbach and religion is to the analysis of fetishism. Both these analyses take place in terms of Marx's materialist dialectic of the form of value.

A fundamental point of Hegel's master/slave dialectic, and one which Marx accepted, is that the categories of master and slave presuppose (mediate or reflect) each other, so that someone is only a master in so far as others relate to him as slaves, while the others only believe that they are slaves because he is a master. Furthermore, the process of mutual recognition is essential to ego development, to the determination of consciousness as human self-consciousness. Hegel rejects as abstract and *a priori* the proclamation of human self-consciousness through Descartes' *cogito sum* or Fichte's *Ich bin Ich*. People constitute their self-consciousness of themselves *as* people through a process of social mediation. Only through the relation to another as a fellow man can one first relate to oneself as a man. The other thereby serves as the form of appearance of the species man.[100]

Marx applies this dialectic of reflexive categories to the form of value, x of A = y of B. To begin with, A and B are natural objects, use values, but not yet commodities because they have not yet been determined as exchange values. For the purposes of eventual exchange, A has been equated with B in a definite proportion. A is then able to relate to itself as a value as a result of its setting itself equal to B *as a value* (as the embodiment of labor, not as a use value, for A and B are not equal as use values since they have different natural properties). Insofar as it relates to itself as a value, A distinguishes itself as a value from itself as a use value. Whereas the use value of an object appears in the natural characteristics of the isolated object, its labor value appears only in its relation to another commodity as qualitatively equal in certain quantities. Value thus only receives a unique form distinct from use value through its presentation as exchange value.

---

[100] Cf. Marx's references to these explicitly Hegelian arguments in footnotes, *ibid.*, S, 18, S. 23.

The simplest form of (labor) value is the *relative* form of the value of A relative to B, where A is seen to embody labor because it sets itself equal to another commodity and the only basis of comparability is the common characteristic of embodying labor. The fact that A reflects its value in the value of B, that A is mediated by B, has, in turn, an effect upon B. B is no longer simply a use value, but is determined as an exchangeable use value. B now possesses the form of immediate exchangeability with other commodities, the form of being an equivalent. This new determination of B as equivalent is a result of the same equation, x of A = y of B, which gave A the relative form of value. B's *equivalent* form of value does not merely determine B to be a value (an embodiment of labor*)*, but a value which in its physical *Gestalt* of being a use value acts as a labor value for other commodities and is thus immediately present as an exchange value. In this description of the equivalent form of value, the word "immediate" indicates that there is no third term, such as money, mediating the exchangeability of B for A. The case where A is sold for money and that money is used to buy B (the formula used later in *Capital*: C-M-C) is the more advanced case of the money form of value and is excluded from the simple relative or immediate equivalent form of value.

Considered purely as value, a commodity consists solely of labor, it is a crystallization or objectification of human labor. However, insofar as the natural form of a commodity reveals the labor which went into producing it – and not all commodities show such physical traces – what it presents is signs of specific, concrete labors (e.g.. weaving*)* and not of abstract human labor as such. To determine a commodity as a value, as the embodiment of abstract labor, requires a process of abstraction. "To take linen as a physical expression of human labor, one must ignore all that really makes it a thing." To consider this process of abstraction a subjective act of a judging subject would be idealism: "Objectivity of human labor, which is itself abstract and lacking quality or content, is necessarily abstract objectivity, *a product of thought.* The woven cotton thereby becomes a *figment of the imagination.*" Marx transfers the Hegelian dictum, that essence must appear, into the material realms: the determinations of commodities must be expressed in their material relations. "For *commodities* are *material objects (Sachen)*. What they are, they must show in their own material *(sachlichen)* relations."[101]

Such materialism *(Sachlichkeit)* is neither a "metaphysical" assumption nor an immediate consequence of Marx's situation in a context of rising capitalism. Its distance from both is due to the mediation of critical reflection. It represents a critique of Hegelian idealism and utopian thought, on the one hand, and that vulgar materialism which takes money as the really real, on the

---

[101] *Ibid.*, S. 17.

other. Marx's materialism includes the thesis that the concept, the ideal and the dollar are all abstractions, to be comprehended and demystified in terms, ultimately, of commonsensical natural objects. Marx thus rejects Hegel's claim that the concept (*Begriff*) can objectify itself without needing an external matter. But Marx's approach here goes significantly further, even laying the basis for a possible critique of Heidegger.

As any hermeneute knows, the interpretation of something *as* something must be founded in the matter itself (*Sache selbst*). Where Heidegger speaks in the anonymous passive voice of the way in which beings "are given as present," Marx not only describes the way commodities actively present themselves, but goes on to show how the apparent activity of commodities is a mask for the underlying action of their producers, human beings. This should become clear in the analysis of fetishism. For now, the comparison with Heidegger can clearly be seen in Marx's summarizing statement: "The natural form of the commodity changes into the form of value. However this quid pro quo brings itself about (*ereignet sich*) . . . only within the relationship of value."[102] Here Marx uses the term, *sich ereignen*, which Heidegger has more recently chosen as the term most appropriate to articulating ontological transformations. *But where Heidegger uses das Ereignis as a "place holder" for transformative processes without committing himself to any content, Marx uses it to express a process of interaction between commodities, thus a social process involving the exchange market. The task of Marx's sharp dialectic in Section 1 of Capital is, in fact, to trace this process of ontological transformation from beings as natural use values to beings as social labor values and to comprehend the change in Being as a fundamentally social process.*

The use value of an object depends directly upon its natural properties. These natural properties express its use value. The use value of a commodity is non-relative, non-abstract and non-quantitative. Labor value is otherwise. It is a measure of abstract human labor, which is not simply identical to the specific concrete labor that went into producing the commodity. In the relation, x of A = y of B, A is contrasted to B as a different kind of use value. Only thereby can the *concrete* labor embodied in B express the *abstract* labor which constitutes the value of A. This is because the non-equivalence of the use values of A and B implies the non-equivalence of the concrete labors in A and B, leaving as a basis of equivalence only the fact that A and B were produced by human labor. B (a use value) becomes the form of appearance of A's labor value because A relates to B as an immediate embodiment of abstract human labor. The labor that went into producing B is not taken as specific, goal-oriented productive activity, but only as a form of objectification or realization of human labor in general.

---

[102] *Capital,* p. 56, S. 37.

The natural form of A can express A's use value, but cannot express its labor value. However, by A relating to B as an equivalent in the form, x of A = y of B, the natural form of B is made to express the labor value of A. In this relation, the concrete labor of B becomes an expression of abstract human labor. Thereby, abstract labor and labor value receive *material expression* in terms of concrete labor and use value. In contrast to the determination of use value and concrete labor, the determination of labor value and abstract labor can only take place within the relations of commodity to commodity, and that means within a social context in which commodities are exchanged. The *theoretical* concepts most central to Marx's critique of capitalist production are thus results of capitalist *praxis* and are shown to be such by Marx's theoretical practice.

The simplest forms of value of commodities, the simple relative and the simple equivalent forms, were given in the single equation which expresses the form of simple exchange, x of A = y of B. Economic science in the narrow sense is concerned with the quantitative relations of exchange value, costs, the ratio of x to y, price. This concern, however, presupposes that B has been determined as an equivalent which can be compared to A quantitatively in terms of a common denominator and (insofar as socially necessary labor time is taken to be at least one determinant of exchange value) economics presupposes that the concrete labor embodied in A and B can be determined as abstract human labor.

As a reflection upon these two presuppositions, Marx's analysis of the *form* of value provides a theoretical basis for the substance of value (abstract labor) and the *magnitude* of value (labor time). But economics is concerned with the whole market of commodities and not with the case of the simple exchange of two commodities alone. Further, the market operates primarily in terms of monetary exchanges (not to mention such complications as stocks, etc.) rather than immediate exchange of commodities. The simple form of value must accordingly be expanded to account for the *universal* equivalence of all commodities with each other, the universal determination of all concrete labor as abstract human labor and the role of money in commodity exchange.

When commodity A enters the market, it sets itself in exchange relations with all other commodities. In addition to x of A = y of B, the following equations are also necessary to express the *expanded form of value*: x of A = u of C, x of A = v of D, x of A = w of E, and so on for all other types of commodities. Here the seemingly accidental relation of the two commodities A and B is replaced by a system of relations of value. It becomes clear in this form that the relations of exchange are determined by the relative magnitude of value, rather than value being determined by exchange. That is, the realm of production has a priority over circulation in that value is first of all a function of productive labor time and not primarily of supply and demand. Furthermore, this system of commodity relations provides a complete material expression for the abstraction of human labor.

In the simple form of value the concrete labor in A is explicitly set equivalent to the concrete labor in B and only implicitly thereby determined as human labor. Any other common factor would be purely coincidental since it is implicitly assumed that examples of commodities could be substituted for A or B which would not have the coincidental, inessential commonality. This implicit assumption is made explicit with the expanded form of value, in which A is related to every possible commodity and thus the labor in A is expressed as every possible kind of productive human labor and is thereby explicitly materially *determined as abstract labor.*

A further step in the development of the form of value is taken by reversing the equations of the expanded form. Clearly, if A can be exchanged for any other commodity than it is also true that any other commodity can be exchanged for A. The expanded equivalent or *universal equivalent form of value,* which states this, can be represented as: y of B = x of A, u of C = x of A, v of D = x of A, w of E = x of A, etc. Here all commodities express their value in the commodity A. Through A as a social expression of value, all commodities differentiate themselves as use values (expressed by their own natural form but not by A's natural form) from themselves as exchange values (exchangeable with A). Transitively, as all exchangeable with A, all commodities relate to each ether as magnitudes of value, as qualitatively equal and quantitatively comparable. Here for the first time all commodities *appear* to each other as values, "Their value is hereby given its appropriate *form of appearance as exchange value.*"[103] That all commodities are (labor) values appears in the fact that they are all exchangeable with A. Their embodied abstract labor is expressed by the concrete labor embodied in A and their (labor) value is measured in terms of the quantities of A with which they are exchangeable.

Through a dialectical materialist analysis of the form of value of commodities, Marx has argued that the *essence* of the value of commodities is labor value and that its *appearance* is exchange value. This form of appearance is at once an expression of the essence and its obfuscation. The half-truth character of exchange value, its socially necessary falsehood, provides the basis for bourgeois ideology, with its doctrine of just exchange, its emphasis on the realm of circulation, its rejection of labor as the source of surplus value, its confusion of the private and the social and its distorted view of the nature of money. Marx's analysis of abstract labor is a striking example of the

---

[103] *Das Kapital,* S. 26.

dialectical and materialist character of his methodology, for he analyses the "universal," human labor, as a relation between material entities (the commodities B, C, D, . . .) and, shows how it is materially expressed (in the universal equivalent, commodity A, which mediates all commodity exchanges by positing itself and all other commodities as objectifications of abstract labor). Rather than rejecting the power and dynamic of the abstract as e.g., non-empirical, Marx comprehends the life of the abstraction by focusing on its material expression. This task automatically subjects the ideal which is glorified by ideology to a critique in terms of its actual social manifestation. The analysis of the form of value also clarifies the role of money in the economy by showing its derivation from the simple form of value.

Money is the result of further transformations of the form of value beyond the form of the universal equivalent. In *Capital*, Marx uses linen as an illustration for commodity A, the universal equivalent. One could just as well use any other commodity as the material expression of labor values: corn, wampum, gold, silver, etc. Money is merely a *symbolic* representation of the material universal equivalent. In the form of paper money or bank balances, money is a "pure" expression of value in the sense that it expresses no use value. Its "purity," however, helps obscure its social roots, its derivation from the simple form of value and its relation to concrete labor, to the organization of social production, to the producers. Money specifies a particular social form of exchange, even as it hides its socio-historical specificity.

Whether the universal equivalent is linen or money, it serves all other commodities as a universal corpus of value, a universal materialization of abstract human labor, a universal form of realization of universal labor. Insofar as all commodities reflect themselves as quantities of value in one and the same equivalent, they all reflect each other as quantities of value. The universal equivalent mediates the relations of exchangeability between all other commodities, so that in a monetary economy linen is not immediately exchangeable with coats the way each is with money: if they are exchanged, the one is sold for money and the other is bought with money.

The determination of a universal equivalent with which all commodities are immediately exchangeable is a social determination, as it must be valid throughout the commodity market: "In the form of *exchange values*, commodities *appear* to each other and *relate* to each other as *values*. Thereby, they also relate themselves to abstract human labor as their *common social substance*. Their *social* relation consists exclusively in being for each other nothing but quantitatively different, qualitatively equal, interchangeable and

[104] *Ibid.*, S. 28.

exchangeable expressions of this, their social substance."[104] The form of exchangeability of commodities, their form of value, is their social form, because it is their value, their exchangeability, which sets them all in social rapport.

To analyze capitalist society, in which value is expressed as money, it is thus necessary to analyze the money form as a form of value and to show that the money form is socially specific by distinguishing it from natural forms (e.g. the use value of gold) and from forms of value in other social formations. This is the task of Marx's critique of the fetishism of commodities.

Fetishism as Appearance and Reality

The form of value given by the equations of exchange between commodities is a form specific to an economy where social production is carried out by private labor. Commodities are products of mutually autonomous private labors, which, however, are interdependent in terms of a developed social division of labor. This interdependence is a function of the material differences between the products as use values. The products are not social values in the immediate sense of products of cooperative work, but are only determined as values through their relations to each other or, derivatively, to a universal equivalent. The commodity form of value is a mediated form of social value, to be contrasted both with non-social pre-capitalist production and post-capitalist social production through cooperative association. The social form of commodities consists in their relations to each other, relations of identical human labor, relations of non-identical concrete labors in their abstractions as abstract labor. This relation of the products of labor to each other as products of abstract human labor defines the social form of labor specific to bourgeois (commodity) production.

One might object that all instances of labor by people in all societies have been instances of the general category (universal or abstraction), human labor. But Marx materialistically insists that the standard of "sociality" whereby the form of labor is categorized must "be drawn out of the nature of the characteristic relations of each mode of production and not out of conceptions external to them."[105] The labor involved in producing a commodity is not immediately determined, as human labor, but rather as concrete labor of weaving, tailoring, etc. Only *mediately*, through the relations of exchange, is such labor determined as abstract labor by being set equal to

---

[105] *Ibid.*, S. 32.

the labor embodied in the universal equivalent. Wage labor, for instance, is abstract labor worth so much per hour when its concrete labor power is posited as relative to gold or wages. Labor, the activity of people in society, is interrelated in capitalist society through a highly developed network of productive specializations. However, this nexus of social relationships has a mediated form: it only comes to expression through the relations of exchange of the products of labor, the commodities. In particular, people's social relations are expressed by the universal equivalent commodity or, as relations of wage labor to capital, interpersonal relations of productions are expressed in monetary terms.

The universal equivalent exhibits a tendency to conceal its origin in commodity relations, let alone in the social relations of the people who produce the commodities. This tendency towards a false appearance (*falsche Schein*) begins with the simple equivalent, B in the equation x of A = y of B. In the simple form of value, A takes the active role, setting itself equal to B. The character of equivalence that B has in the simple relation only as a result of this reflexive relation *seems* to belong to B "naturally." It seems that A relates to B *because* B is a materialization of abstract human labor, something already present as a corpus of value. The illusion is, however, not yet complete in the simple form, because along with the tendency to attribute B's immediate exchangeability to B as a physical characteristic independent of its relation to A is the opposing tendency to treat A and B as interchangeable by reversing the equation.

With the development of the form of value from the simple equivalent to the universal equivalent, the tendency for the underlying relations to be obscured is strengthened. The universal equivalent seems merely to be a "natural" expression of value because the opposed moments of this form of value no longer develop identically for the commodities which are related. The form is now an asymmetric many-to-one relation of all possible types of commodities to the single equivalent. This form distinguishes the universal equivalent as something entirely set apart from all other commodities. Further, the character of the universal equivalent is in fact no longer dependent upon its relation to any one other commodity and it therefore seems to be prior to and independent of each of the relations which determine it. The tendency of the universal equivalent to obscure its origin in the form of value and its nature as an expression of exchange relations and of the social relations which underlie exchange resulted in the myriad confused theories of money which Marx confronted.

The system of commodity production, in which money plays a well-defined role, is one possible means for relating the various specialized labors which take place in a technological society. People here relate their various labors to each other *as* abstract human labor *by means of* relating their products to each other *as* values. The social, inter-personal relation is concealed by the physical form of the commodities. In commodities, the substance of value (human labor) is no longer visible. In order to relate their products as commodities, people are forced to equate their labors with abstract human labor. The relations of commodities, expressed in money as the universal equivalent of all commodities, force this upon people regardless of what theories people may have concocted to explain the form of their activity. In a society of commodity production, the economic relations, which are after all results of human interaction, take over a priority and autonomy from people's conscious intentions. In such a situation, economics forms a "base" underlying the "superstructure" of all attempts to comprehend and control economic forces.

Marx's theory of capitalist society uncovers the determination of the magnitude of value by labor time as the secret hiding beneath the apparent development of the relative value of the commodity. It does this by showing how the form in which value appears in capitalist production forces people to equate their labors to abstract labor by equating all the products of labor to a universal equivalent which expresses the objectification of abstract labor. It is precisely this form of value which conceals the social relations of the private laborers and the social specificities of the private labors in objects rather than revealing them. The form of value which the product of labor takes in the historically-specific bourgeois mode of production, the exchange value of commodities, results in a fetishism of the products as things independent of social relations.

The mode of production analyzed in *Capital* is one in which private producers establish social contact with one another through their private products, through material objects. The determinations of social criteria for production, consumption and distribution are mediated by commodity relations. Consequently, rather than the social relations of products *being* and *appearing* as social relations of people at work, they appear as material relations of people or social relations of material objects.[106] The social relations of people make their appearance in the realm of material objects in a form which seems independent of the interpersonal realm. In analogy with religion, in which, according to Feuerbachian critique, social relations of people are worshipped as sacred entities in a heavenly realm prior to human society, Marx calls this apparent transference "fetishism."

Marx goes beyond Feuerbach's critique of religion by situating contemporary religious forms within the superstructure of commodity production. He

---

[106] *Ibid.*, S. 40.

argues that, "for a society of commodity producers, whose general social relations of production consist in relating to their products as commodities, thus as *values*, and in relating their private labors to each other in this *material* form as *identical human labor*, *Christianity*, with its cult of abstract man – namely in its bourgeois development, Protestantism, deism, etc. – is the corresponding *form of religion*." Against Feuerbach, Marx insists that an intellectual criticism of religion, especially an argument which itself adopts a cult of abstract man and which fails to situate religion in a social context, cannot be effective. As early as in his fourth "Thesis on Feuerbach" Marx had pointed out that a prerequisite to defeating religion was the removal of its original social motivation.

In *Capital*, Marx has precisely located the origin of religious fetishism along with fetishism of commodities in the cult of abstract man and abstract labor, which mediate social relations under capitalism. Religion cannot be abolished without eliminating commodity fetishism. The latter can only disappear once the form of value is transformed into a social structure which does not obscure the social, interpersonal relations of production. This is tantamount to a call for the overthrow of wage-labor and commodity production. It should, incidentally, be clear that Marx's position is forcefully opposed to any cult of abstract man, including the "humanism" Heidegger tries to saddle him with. *Capital*, Marx's mature theory, which was at least implicit in his earlier *Manuscripts*, does not object to alienation on the basis of some idealistic conception of the nature of man. It reveals the process whereby the social relationships, which in every society determine the form of human potential, are alienated from people through the fetishism of commodities. The economic basis for this development is given by the form of value of commodities as exchange value. This form of value is due to the relation of the worker as wage laborer to the capitalist as owner of the produced commodities.

Although abstract human labor is the substance or *essence of value* in bourgeois society, the *form of value* taken by this essence, the form in which it manifests itself in appearance, is *exchange value*. This appearance is not an arbitrary lie, but a socially necessitated illusion. As part of his analysis of the essence of (labor) value, Marx was thus obliged to analyze the form of material expression of this value in the exchange value of inter-related commodities. This analysis of the form of material expression of value, was, in fact, Marx's major contribution to the theory of labor value, combining classical political economy's concern with the appearance of exchange value and proletarian ideology's insistence on the essence of human labor value.

Such a unity of theory and practice within the concerns of social theory is necessary if social theory is to contribute effectively to the transformation of capitalist society. Social reform is objectively impossible under advanced capitalist society. The contradictions of capitalism do not make reformation easier, rather, they infect any attempts at change, distorting and strangling them. Surely any action against the social system which unknowingly adopts that very system's ideologies (e.g. a naively moralistic demand for equality based on the cult of abstract man) without subjecting them to social critique is doomed from the start to fail in its simplistic objectives. The failure of reforms is today attributed to "cooptation" and not to a lack of theory, however.

Cooptation can, nevertheless, be comprehended within Marx's theory of capitalist society. A reform is undertaken on the basis of its inherent or essential qualities as perceived abstractly. When the reform is put into practice, however, it is mediated by its social context, by all the other existing criteria which also condition its object. The aided reformist criteria interact with the pre-existing social criteria to develop qualitatively new and unexpected effects. Thereby, something which may "work in theory" may not have the anticipated results in practice. This is particularly true in the context of bourgeois social relations, where essential determinants are obscured and distorted. The appearance created by the effects upon reforms due to pervasive commodity relations and the consequent obfuscation of commodity fetishism tend to result in cooptation. The attempted reform is transformed under the commodity form of relations. The relation of the conscious reformers to their intended reform is obscured and takes on the apparent form of a relation of "the owners of capital" to a reform which has been altered to serve their interests. The apparent benefactors of the distorted reform need not be identifiable, they need not exist as individual people. The reform has simply been made to serve the criteria of exchange value or capital and the human agents presumed to be behind this fetish are invisible, autonomous, abstract, indefinable: the System, the Man, the They, the Power Elite.

Abstract moralism and purely practical power politics fail equally to challenge the status quo if they do not reflect upon the ways in which reforms are mediated by their form of material expression in the given social context. Because the various determinations interact in opposing ways and develop beyond their original forms, the materialistic analysis of the form of appearance or the essence must be dialectical. It must, furthermore, be informed by an historical sense, both of the continuous social dynamic and of differing historical forms as various possible forms of appearance of a particular essence.

Idealism and empiricism, theory without concern for practical expression and the acceptance of appearance without comparison to essence are both ideological, uncritical. They are instances of the one-dimensional thought which corresponds to bourgeois society without being able to transcend it.[107] Marx's presentation of an analysis of capitalist society in *Capital*, on the contrary, unifies the theoretical and the practical within theory, thereby penetrating for any political practice which it informs the illusions generated by the fetishism of commodities.

---

[107] Cf. Herbert Marcuse, *One-Dimensional Man* (Boston: Beacon, 1964).

PART III. MARTIN HEIDEGGER: META-ONTOLOGY AS INTERPRETATION AND TRANSFORMATION OF THE WORLD

Martin Heidegger thinking about transformations of Being.

Transitional Remarks

Marx's theory is intimately related to the social context of its times: to the establishment of the power of the bourgeoisie and the rise of an industrial proletariat, to the ensuing class conflicts and the concomitant crises of the capitalist economy. Heidegger's thought also corresponds to his time, in particular to the technological advances surrounding World War II and the accompanying technification of social existence. The significant developments of the century which separates Marx from Heidegger, not least of all in the form of the capitalist economic system, raises the question of Marx's relevance to our world. Has society changed so drastically qualitatively and essentially that Marx's categories no longer get to the root of the matter and that his critique is no longer fundamental? Is Heidegger for this reason justified in ignoring commodity relations as such in favor of a thorough and historical analysis of technological rationality, reversing the relation of priority between what Marx occasionally referred to as base and superstructure? *Dialectic of Enlightenment* by Horkheimer and Adorno has demonstrated that it is possible to comprehend within a Marxian framework the development of technological rationality from the early Greeks to German fascism and American pop culture. Heidegger, however, rejects such an approach, insisting on the need for a new set of essential categories, in fact, for a new language and a new comportment of investigation.

The following chapters will present an interpretation of Heidegger's central path of thought in his mature work. They will investigate how Heidegger's system stands up to the standard set by Marx's dialectical materialist method and the explanatory power of his theory. Guiding the inquiry will be Adorno's claim that Heidegger hypostatizes Being as divorced from beings. Whatever the conclusion, it cannot be considered final, because Heidegger's important discussions of, for instance, language and the ontological difference must remain beyond the present scope. However, where such subsidiary topics contribute significantly to the problematic here, they should be expected to reappear within Heidegger's central argument.

The path of thought, according to Heidegger, is the questioning of Being. For the sake of a preliminary orientation, what Heidegger means by "Being" can be roughly approximated in a series of steps and through comparison with parts of Marx's system. Being is the universal determination of beings as beings. In various historical periods this determination has taken different forms so that *all* beings were present *as*, for instance, creations of God, objects for subjects or materials for labor. Being can thus, on first approximation, be considered the prevailing world-view of an historical era, conceived, however, on an ontological rather than political or aesthetic level as an interpretation of all beings as such in general. But talk of world-views is too subjective, leaving the decision on how to interpret beings up to the individual's wager.

Already in *Being and Time* (*Sein und Zeit*, hereafter referred to as *SuZ*) Being is treated as an *a priori*: prior to subjective perception, beings are always already *given* as interpreted. But in *SuZ*, as part of its man-centered fundamental ontology, the objectivity of Being is reduced to the subject's individual network of meanings, remaining subjective and individualistic even if prior to conscious choice. Even in *SuZ*, however, there is a tendency, developed in Heidegger's later writings, to talk of man as "being there" in a "clearing of Being." This circumstance is prior to the hermeneutic "as" and can perhaps be construed, as follows: beings are present to people and as present are given with certain meanings. While these *meanings* are to be attributed to the subject's manifold of signification, the *presence* of the beings is independent of this subjective hermeneutic sieve, prior to it. "Being" applies to this latter level, as the determination of the character of the presence of beings as given beings in general, not as the determination of the meaning of individual beings or even of the system of their possible meanings. The question of Being is a reflection upon presence.

Heidegger represents his task as retrospective first philosophy, which tries to uncover the fundamentals through systematic and historic back-tracking. But the important motivation of his project lies elsewhere. The question of Being would be an academic matter if Being were immutable. But Being is subject to historic variation and its historic forms overlap at any given time. The questioning of Being is propelled by a yearning for a new epoch of Being, whose coming entails sorting out the contemporary forms of Being, which obscure their own history and interconnections. Of special concern is the reified form of Being which is associated with our technological epoch. In *SuZ* this form is criticized as inadequate under the category of presence-to-hand in contrast to readiness-to-hand or the ecstatic temporal structure of human existence. Later, technological "stock" is opposed to a dynamic conception of the thing.

Marx's thought can also be understood in terms of this question of Being. Marx is concerned with beings insofar as they are products of human labor. Thus it is clear that for him beings are determined by human labor in accordance with human intentions, needs and capabilities. This differs little from Heidegger's discussion of the manifold of meanings, which focused on the pre-capitalist workshop to develop the non-capitalist concept of readiness-to-hand in relation to the individual's project, meaning structure and interpersonal relations. The difference is that Marx's analysis – in addition to and related to having more concrete social content – is socio-historically specific. In moving now to a comparison of Marx and Heidegger on the presence of beings as such, rather than on the determination of individual beings, both the similarity and the contrast remain. As socially produced, beings have a general character of presence which corresponds to the socially prevalent mode of production, to the unity of the technical forces and social relation of production. *The epochal history of Being corresponds to that of social production.* Further, the important motivation behind Marx's analysis lies in the contemporary contradiction between two ontological forms: use value and exchange value. The point is to reconcile these through a "social value" form of beings. Because Being corresponds to the mode of production, the reign of social value presupposes the rearrangement of the forces of production developed under capitalism (the industrial revolution and subsequent technology) in accordance with transformed, post-capitalist social relations.

Marx's work can be understood as an ontological investigation and thus as an alternative to Heidegger. The form of value, whose analysis is so central to *Capital,* is an ontological category, a determination of the Being of commodities, of their form of presence. Commodities are *given as* use values and *as* labor values, as embodiments of useful labor and of abstract labor. This historical ontological determination is social. It has a base in economic structure and economic history, as Marx shows in his sharp dialectic of the form of value, in his entire analysis of the real process of exchange (in production, circulation and the total economic process) and in his retrospective history of property relations, whose perspective was designed to show the historic determination of the commodity form.

The methodology of Marx's critical social theory as hermeneutic ontology provides a point of reference in the investigation of the Heideggerian alternative. Of particular importance is the relation of Being to beings, which in Marx takes the form of the relation of the abstract to the concrete, as mediated by historically-specific concepts. In the *Grundrisse,* general (trans-historical) concepts are abstractions from the mere concrete historically-specific concepts. In *Capital,* on the other hand, historically-specific exchange value is derivative of the essential labor value, being merely one form of the appearance of this abstract essence. The mediation of these opposed relations takes place through Marx's two-way trip from the concrete to the abstract and then back to the concrete, but now criticized as limited in terms of the abstract potentials. Marx's abstractions, which are necessary for a critical stance – for distinguishing second nature from the natural and forms of appearance from the essential – are themselves arrived at through the concrete appearances, which are thereby critiqued. Without resorting to arbitrary or external metaphysical frameworks of interpretation, Marx satisfies the need for abstract concepts, trans-historical categories and essential characteristics in accordance with the hermeneutic principle of basing interpretation upon the subject matter itself, explicating what is most appropriate to it.

The following chapters trace Heidegger's attempt to uncover the Being of beings through appropriate interpretation of beings. Heidegger's thought is followed through three lectures which present the development of his mature system, emphasizing his response to the problematic just discussed in terms of Marx. This problematic is unavoidable for any truly post-Hegelian, historical, hermeneutic ontology.

The author in 2002, visiting the hut where *Being and Time* was written.

Chapter VII: The Work

The Art Work and History

In Germany of the mid-1930's, with the life-and-death struggle between Hitler's National Socialists and an active communist movement, the question of how the character of future society would be determined was no academic matter. Taken within this context, Heidegger's 1935 essay on art can be interpreted as a philosophic approach to the burning issue. Heidegger does not assume that history is determined by political world-views which are backed by political power, but raises the ontological question in its historical form: How is the character of a given historical era determined; specifically, how was ours determined and how is the coming era already being determined? Heidegger phrases his approach to this problem in terms of art: Is art an "origin" of historical ontological change? Comprehension of his answer to this question presupposes a careful reading of Heidegger's purposefully ambiguous essay on the origin of the work of art.

While no interpretation of Heidegger's thought can ignore *Being and Time*, any final evaluation must be based on his subsequent reversal of the position stated there. In *SuZ* Heidegger founds the revelation of being in the labor of man, human *praxis*: the involved Being-in-the-world of *Dasein*. This relationship of founding was mediated by a concept of truth as discoveredness. "Being (not beings) are given only insofar as truth is. And truth only *is* if and when *Dasein* is."[108] The notion of fundamental ontology as the analytic of *Dasein* seemed to be a primary methodological principle of Heidegger's ontology.

However, from the perspective of Heidegger's later writings – the perspective upon which our considerations will be based – the analysis of *Dasein* provides no foundation for building a philosophical system in the sense of Descartes' *cogito sum*, but rather a point of departure for a path of thought along which such preliminary relationships must be reinterpreted and reversed as part of a

---

[108] *Being and Time*, §44. p. 272. Cf. *Sein und Zeit*, S. 230.

reversal of thought. This reversal takes place repeatedly on many levels: in the difference between the tradition of metaphysics and Heidegger's own thought, through the development from *Being and Time* (1927) to *Time and Being* (1962), foreshadowed in the early distinction between inauthentic and authentic, by the twists in dialectical sentences, thanks to the ambiguity of phrases betokening possession and by means of the reinterpretation of individual concepts.

The reversal of thought can be followed effectively in terms of Heidegger's lecture, *The Origin of the Work of Art*. This work, in fact a pivotal point in the author's *ouvre*, makes a well-considered break with metaphysics by questioning Hegel's seminal philosophy of art, rejecting the subjectivism of traditional aesthetics and deserting the royal road to knowledge in favor of lingering along the overgrown and forgotten dead-end trails, the intellectual *Holzwege*. In the reversal of the meaning of the lecture's title, the origin *of* the work, or of the equivalent phrase, the setting-itself-into-work *of* truth, the subjectivistic standpoint – not yet sufficiently overcome in *SuZ*, where *Dasein* is still a foundation – is surpassed. We shall take this specific example of Heidegger's reversal as a model for concretely understanding the abstract formulations in which Heidegger presents his final position.

The traditional characterization of the work of art is in terms of a person, either the creator or the audience, for whom the work is a representation, a second presentation, of something that was already present, revealed, discovered for someone as an object or idea. In the language of *SuZ*, "Revelation is an essential mode of being of *Dasein*. . . . Beings are only *then* discovered and only revealed *as long as Dasein* is at all."[109] Because discovery or revelation is commonly taken to be essentially a function of man, an art work can only present what its human creator has discovered and its human audience rediscovers. As such, a work of art can *transmit* truth, but nothing more; the origin of the work of art as a conveyor of truth is then its creator, the artist, human subjectivity: *Dasein*.

---

[109] *Ibid.*, p. 269, S. 226.

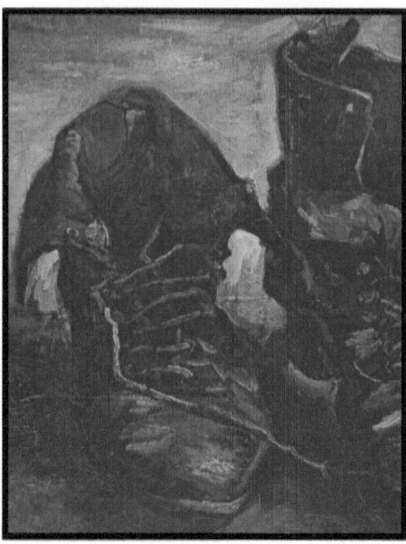

Heidegger's first step in his analysis of the origin of the work of art is to move away from the traditional characterization by citing an experience in which a work reveals truth independently:

> Van Gogh's painting is an opening-up of that which the tool, the pair of farmer's boots, in truth is. This being moves into the unconcealment of its being. . . . There is a happening of truth at work in the work, if an opening-up of the being takes place there into that which and that how it is.[110]

Somehow, claims Heidegger, the oil painting of the peasant woman's shoes itself reveals the nature of the shoes as serviceable and reliable tools in the farmer's world. Such a revelation is often reserved for philosophical contemplation and even *SuZ* restricts it to the understanding of *Dasein*; but in the preceding quotation it appears to be the work of an art work. Art is then characterized by Heidegger as the setting-into-work of the truth of the being, where the "setting" cannot be understood as a human act of placing already understood truth into an artful representation, but rather the essential function of the work itself is to discover on its own and to present for the first time the truth about something.

---

[110] Martin Heidegger, "The origin of the work of art," in *Philosophies of Art and Beauty*, ed. Hofstadter & Kuhns (New York: Modern Library, 1964), p. 664f. Cf. Martin Heidegger, "Der Unsprung des Kunstwerkes" in *Holzwege* (Frankfurt: Klostermann, 1963), S. 25.

What we have paraphrased as "presenting" the truth about something, Heidegger more precisely names an opening-up of the being in what and how it is. This phrase refers back to the origin in the essay's first sentence as that from which and through which something is what and how it is. The title of the essay cannot be understood according to the traditional notion of art, where the artist is the origin of the work, but must already be seen to point to the origin-character of the work itself: the work of art as origin of truth.

Van Gogh's painting evokes a whole world of relationships in which the shoes which it pictured must, as tools, stand. Heidegger then applies this notion of a world of relationships to the work itself and asks how a work of art is situated within *its* world. Is the painting to be considered in terms of its surroundings in a museum, in its creator's biography and *ouvre*, or in its social and historical circumstances? Heidegger's answer is that the world of the work is not independently present, but is itself determined or revealed by the work. The work belongs solely in the region which is opened up by the work itself. This is because the "working" of a work takes place only in such opening up.

In revealing the truth about some being, a work is self-revelatory, revealing itself and opening up a realm for itself which it also reveals. All art is, one could say, dramatic, opening a stage within, yet separated from mundane reality, a space and a framework of significance in which the truth which the work reveals can stand out obtrusively. The work of art is the origin of stage, script and message, of its world and itself as well as its "object." It sets itself into the work of presenting truth about something which it is not.

Even beyond its self-revelatory character, the work of art is the revelation of the world of an historical people, says Heidegger:

> The temple as an architectural work joins and gathers around itself the unity of those paths and relations in which birth and death, illness and blessing, victory and disgrace, endurance and decay win the form and the run of human nature in its destiny. The persisting breadth of these open relations is the world of this historical people. From and through them it first comes to itself and back to the completion of its determinations.[111]

To determine what is meant by this function of art is the central problem of our investigation. What revelation of the truth of the being or of being meant in Heidegger's pronouncements is far from clear, and the claim that this is the essential function of art must remain unsupported until this is clarified.

---

[111] *Ibid.*, p. 669, S. 31.

Whether the reversal of thought has at least been suggested in its full scope or whether only the first steps have been noted has yet to be determined. Perhaps most urgent is to interpret the significance of the work of art as origin of the world of an historical people. If a temple gathers together and joins for the first time the dimensions of history, has it then created the unity of the times or merely reflected it? Is the revelation of an historical world merely the re-presentation – perhaps for the first time making conscious – of what already exists and determines the life of man whether he is aware of it or not? *Or are the historical alternatives formed and selected in works of art*, among other ways, so that the work of art is the origin of the world we live in as well as the work's own world? This problem poses the crucial question of Heidegger's thinking about Being *in nuce*. There may be no better way of understanding his position than to interpret carefully what he has to say about the relationship of art to history.

Art and Being

At the close of the preceding section, the possibility was proposed that the work of art may in some way represent prevailing relationships without in any manner changing the structure of reality. An art work would then provide a unity in accordance with its aesthetic form for the material which is historically given independently of the work. The realm of art would be a medium for the transmission of interpretations of the world, but not a creative source of new interpretations.

This possibility seems, however, to be immediately rejected by Heidegger. The Greek tragedy, for instance, was no simple allegorical representation of the battles of the gods or of the moral cosmic order by actors on the stage. Pre-existing gods were not described nor were ethical values indoctrinated; rather, the gods were brought into existence and the values determined in the drama itself; they were created and formed there, to be preserved in the language and tradition of tragedy.

> In the tragedy, nothing is presented, rather the battle of the new gods against the old is waged. When the linguistic work appears in the speech of a people, it does not talk about this battle, but transforms the speech of the people so that now every essential word wages this battle and places in question

---

[112] *Ibid.*, p. 670, S. 32.

what is holy and what is unholy, what large or small, courageous or weak, exalted or fickle, master or slave.[112]

The artistic presentation cannot be separated from its creative and formative functions: Zeus cannot appear on stage as a nondescript pawn in the action, but can only *be* Zeus *as* holy, grand, courageous, exalted and masterly.

When a temple is erected to Zeus, the erecting includes consecrating, which is declaring the holy as holy. Consecration means making holy in the sense that in the erecting of the work the holy is opened up as holy and God is called into the openness of its presence. Just as *Dasein*'s understanding always reveals beings as already interpreted, so the work of art is necessarily an interpretation of that which it presents.

Dasein, according to *SuZ*, reveals and relates to beings as already interpreted within a framework of meanings called its world. This world is a part of *Dasein*'s existential structure, a hermeneutic sieve for relating to other beings, not the totality of these other beings themselves. Heidegger analyses the being of a work of art much as he analyzed *Dasein* – as a being which, unlike beings present-at-hand and ready-to-hand, erects a world in which truth is opened up and beings are revealed: to be a work means to erect a world. Naturally the ontological analyses of man and art are not identical, the former having an ecstatic care structure and the latter merely the structure of the setting-itself-into-work of truth; however, they both have a world, in the one case disclosed, in the other erected.

The question now is: If the world of *Dasein* is part of his structure and that of the work of art part of its, then do these two worlds have any relation to each other? What did the world of the tragedy have to do with the Greek's world and what does the world of van Gogh's canvas have to do with our world? We have already heard that tragedy raised into question what is holy and unholy, what large and small, etc. Presumably the decisions reached by the tragedy are important for us.

In connection with the work's erecting of a world, for instance, Heidegger says, "Where the essential decisions of our history come to pass, are taken over by us and left behind, forgotten and again sought, there the world is worlding."[113] Our history, the history of mankind, is thus viewed in terms of crucial changes in our interpretive framework, in our world. One way in which these changes are brought about is through the work of art. What the world of the work undergoes in the way of decisive changes is transmitted to the world of people when it is taken over by people in their relationship to

---

[113] *Ibid.*, p. 671, S. 33.

the work. In the role of preservers of the truth which a work reveals, people join each other and the work in a common historical process.

> For the preservation of the work does not isolate men with their individual experiences, but pulls them into a relation of belonging to the truth which happens in the work. Thereby, men's being for and with one another is grounded as the historical exstasis of *Dasein* out of relation to unconcealment.[114]

The character of historical decisions is now somewhat clarified. History, for Heidegger, consists in changes in the relationship of man to the unconcealment of beings. This does not contradict the talk about changes in the world, but rather explains the possibility of changes in the framework of the meaning of beings in terms of the revealing and concealing of changing aspects of beings. The effectiveness of the work does not, according to Heidegger, consist in an effect. Rather, it rests in a change which arises out of the work, a change, however, of the unconcealment of beings, i.e. a change of Being.[115]

The work's relationship to the interpretation of beings motivated Heidegger to say that the essence of art is poetry, for "language is the house of Being" in the sense that the interpretation of something as something, the *hermeneutic as*, is fundamentally linguistic, even in its non-thematized, pre-predicative stage (cf. *SuZ* §32). Language is thus the central dimension of history for Heidegger. His historical reconstructions trace the translations of ideas from language to language (Greek, Latin and various stages of German) and from writer to writer (Parmenides, Plato, Aristotle, Hegel, Nietzsche), while his own struggle to reverse the history of thought is fought in terms of reinterpreting grammar, phrases and key terms of metaphysics.

The ability of art to set truth into work is now formulated as follows: truth as the clearing and concealing of beings takes place as poetry. Art has to do with the revealing and concealing of beings because art, as essentially poetic, is linguistic in nature. Here language is not understood by Heidegger as a system of spoken or written symbols which express attributes and relations of already present beings. Rather, language not only further determines what is already manifest or concealed as *so* interpreted by means of words and sentences, but language brings the being as a being into the open in the first place. What poetry does linguistically to open up a world in which beings are named as beings, as so and so interpreted, every work of art does in its own

---

[114] *Ibid.*, p. 690, S. 55f.

[115] *Ibid.*, p. 693, S. 59.

way. The work's interpretations are creative, original. They replace the mundane clichés of the familiar world and thereby bring beings obtrusively into focus in a new way. Due to the poetic nature of art, it happens that the work of art breaks out an open place in the midst of beings, a place in whose openness everything is different from everywhere else. By clearing an opening for a new interpretation of beings, art plays a key role in the movement of history. The harmonic order of Greek art, the religiosity of medieval works and the subjectivistic perspective of art since the Renaissance can be seen in relation to the prevailing interpretations of beings throughout history:

> Whenever beings as a whole, as beings, require grounding in openness, art attains to its historical essence as institution. This happened in the West for the first time in Greece. What was to be called Being was definitively set into work. The beings as a whole which were thereby revealed were then changed into beings in the sense of creations of God. This happened in the Middle Ages. These beings were again changed at the start of and during the modern age. Beings became calculably controllable and transparent objects. Each time, a new and essential world broke out. Each time, the openness of beings had to be erected through the fixation of truth in the *Gestalt*, in the beings themselves. Each time, the unconcealment of beings happened. The unconcealment set itself to work and art accomplished this setting.[116]

Thus far we have given Heidegger a "nominalist" interpretation. Being is nothing beyond the characteristic of individual beings as interpreted as beings. Being thus changes when the totality of beings are differently interpreted within the world shared by people and works of art. In the classical age beings were beings as parts of an harmoniously ordered cosmos; in the Middle Ages as creations of God; and recently as calculable material. Historical change is produced by beings themselves, such as works of art, not by external forces. Such a nominalist interpretation which gives beings a priority over Being, is a plausible one for all of Heidegger's writings. This may be due to the hermetic principle of phenomenology, "to the things themselves," which is repeatedly referred to and adhered to in Heidegger's analyses. Where *SuZ*'s search for the wholeness of *Dasein* confined all within *Dasein*'s own existential structures (Being-in-the-world. Being-with, temporality), the analysis of the work of art tries to retain earth and world, creator and preserver, beings and truth firmly within the art work's internal structure as the setting-itself-into-work of truth.

---

[116] *Ibid.*, p. 697, S. 63f.

The nominalist interpretation is, however, problematic. Most of the ontological structures of a being point to other beings and, even overlooking this, Heidegger's attempt repeatedly fails, as he admits in the midst of his essay on art: "The attempt to determine the work-character of the work purely out of itself has shown itself to be impossible."[117] But certainly the crucial question is how to understand a history shared by all beings in terms of an ontology of monads without a god to insure harmony. How is it, that is, that the manifold worlds of all people and all art works in a given historical period share a single interpretation of beings, a common meaning of Being? This question demands a second look at Heidegger's essay on art.

The Primacy of Being

The previous chapter's interpretation was a desperate attempt to rescue Heidegger from a mystical position. In line with the nominalist tendency, everything that smacked of metaphysics, that referred to something that cannot be grasped in the hands and reckoned with, was argued away at the expense of any ability to understand history. Whatever value this approach to Heidegger may have, and whatever it has in fact revealed of his position, it cannot be considered the whole story, even in outline form. On the basis of such an interpretation we will never be able to explain the crucial fact that in each historical age there is a prevailing meaning of Being. To fail to see Heidegger's struggle with this question would be to miss the whole thrust of his thought.

Heidegger approaches his question of Being and beings by way of a reflection on truth. *SuZ* argued that propositional truth is only possible if the beings denoted have been revealed beforehand. This prerequisite revelation was considered in terms of human *Dasein*'s Being-in-the-world. Associating revelation with *Dasein*, *SuZ* claimed that truth *is* only in so far and as long as *Dasein* is. However, in discussing art and truth, Heidegger shows that this revelation can also be a function of the work of art as the setting-into-work of truth, where truth is taken in its primordial sense of revelation, precondition for propositional truth. Whereas in *SuZ* Heidegger stressed the role of *Dasein* in the revelation of truth, in *The Origin of the Work of Art* he underlines the passivity or receptivity of *Dasein* in this role: "But it is not we who presuppose the unconcealment of beings, rather the unconcealment of beings determines us in such an essential way that we are always placed after

---

[117] *Ibid.*, p. 682, S. 46; Cf. p. 667, S. 29.

unconcealment in our conceptions. Not only that toward which knowledge is directed must already be somehow unconcealed, but also the entire region in which this directedness toward something moves and also that for which an adequation of a sentence to the subject matter is manifested must already take place as a whole within unconcealment."[118]

A reversal in thinking has switched the significance of the fact that propositional truth requires the unconcealment of its object and of that object's world from: propositional truth presupposes *Dasein*, to: *Dasein*, in order to know anything, presupposes primordial truth as unconcealment of beings. Extended to the work of art, this means that the work does not create its world in erecting it, but the work presupposes an already given world. Furthermore, to truth as unconcealment belongs its battle with concealment. The clearing of Being which is constitutive of *Dasein* and of the work is determined by an unconcealment, which is in turn characterized by a self-concealing concealment. "Concealment conceals and displaces itself. That is, the open place in the midst of beings, the clearing, is never an empty stage with continuously parted curtain on which the play of beings plays itself out. Rather, clearing happens as this double concealment. Unconcealment of beings is never a merely present circumstance, but a happening."[119]

This processual notion of truth as the interplay of concealment and unconcealment is absolutely central to Heidegger's thought, from his conception of phenomenology (cf. §7 of *SuZ*) to that of the *Ereignis* hinted at in his mature essay on *Time and Being*. The relationship of such truth to the work of art, theme of the essay on art, is an exemplary analysis of the relationship of Being to beings. The kernel of the analysis is presented in a single paragraph, worth quoting at some length.

> Truth is the primordial battle, in which the open is fought for in one way or another, in which everything stands and out of which everything holds itself back that shows itself and erects. . . . The openness of this which is open (truth) can only be what it is, namely this openness, if and as long as it arranges itself in its opening. Therefore there must always be a being in this opening, in which the openness can take its stand and gain fixity. . . . With the reference to the openness arranging itself in the opening, thought reaches a realm that cannot yet be analyzed here. Only this is remarked: that if the essence of unconcealment of beings belongs in any way to Being-itself (cf. *SuZ* §44), then the

---

[118] *Ibid.*, p. 677, S. 41.

[119] *Ibid.*, p. 678, S. 42.

> latter on the basis of its essence lets the play-space of
> openness (the clearing of the there) happen and brings to
> pass that in which every being comes to pass in its own way.[120]

Here we can distinguish two basic theses on the relationship of Being to beings. The thesis extensively thematized in this section of the essay is: openness (world, truth, clearing, Being) can only be what it is if and when it is arranged in a being within the opening. Taken together with the previous premise about primordial truth – that for a being to be discovered presupposes an openness – this thesis implies a dialectical relationship between Being and beings, in which each both "presupposes" and "determines" the other. The second thesis is only touched upon in the art essay, to be developed in later writings,[121] namely: Being-itself has a priority; out of its own essence – i.e. unaffected by beings – it opens the clearing in which beings can appear. In this second thesis, the dialectical relationship is largely lost in favor of a linear chain of command from Being-itself (the *Ereignis*) to Being (as the Being of beings), to beings.

These various theses can be ordered as successive interpretations of the essay's title which follow Heidegger's reversal of thought. "The origin of the work of art" can be understood in the following ways:

1. The artist, in forming matter into the work, and the audience, in experiencing the work's psychological effects, are its origin.

2. The work is an origin, namely of the truth about its object.

3. The work is its own origin as that of its world.

4. The work and the clearing in which it stands are – dialectically – each other's origin.

5. Being-itself is the origin of the work in which it sets itself into work.

The remainder of our analysis of the essay on art will be concerned explicitly with the last two of these theses, the others representing the criticized philosophical positions of commonsensical approaches and previous thinkers.

The truth, Heidegger says, is not written in the stars to begin with and then concretized, objectified, displayed, represented, made visible, brought down to earth in the form of an art work. Primordial truth as a clearing and the

---

[120] *Ibid.*, p. 684, S. 49.

[121] Cf. the later edition of Heidegger's essay, *Der Ursprung des Kunstwerkes* (Stuttgart: Reclam, 1960), S. 99, and Martin Heidegger, "Zeit und Sein" in *Zur Sache des Denkens* (Tübingen: Niemayer, 1969), p. 43, S. 46.

work as a being in the clearing go together like two sides of a coin. Clearing of openness and erecting in the opening belong together; they are both the essence of the happening of truth. The emphasis here upon the mutual dependence of truth and work as presuppositions of each other is shifted to an analysis of truth alone as a process, where the clearing and the work are both aspects of its essence.

With the incorporation of the work within the structure of truth, the work becomes a merely formal requirement for truth which has no effect on its content: "Because it belongs to the essence of truth to erect itself in beings in order to bring truth into work, therefore the *relation to the work* lies in the essence of truth as a special possibility of truth to come into work in the midst of beings."[122] The phrase, "relation to the work," was italicized in the 1960 edition of the essay to emphasize that the work was to be considered in terms of the structure of truth (not unlike the way *SuZ* considered objects in terms of *Dasein*'s existential structure with its inherent relations to objects through its world) and that what came to be among beings was truth itself and not merely another being, perhaps influenced by but distinct from truth.[123]

It almost seems as if truth created the work in order to come into existence itself. Heidegger's definition of creating is indeed an extreme anti-anthropocentrism: "Arranging truth in the work is the bringing forth of a particular being, which never was before and never will be afterwards. This bringing forth places the being in the opening in such a way that that which is to be brought is the first to clear the openness of the opening into which it comes forth. Where the bringing forth itself brings the openness of beings (truth), that which is brought forth is a work. Such bringing forth is creating."[124]

It still sounds like a dialectical formulation when something clears the opening it requires in the very act of filling it. However, as Heidegger is quick to add, what seems like pulling oneself up by the bootstraps is in truth flowing along with the tide. The force of the dialectic is annihilated, although the form is retained, because the outcome is determined in advance – if not externally, at least one-sidedly. The being does not create its Being, the work its truth, in bringing it forth; rather, as this bringing, it is more a receiving and accepting within the relation to unconcealment. Unconcealment, truth in the primordial sense, determines the character of the clearing and of the beings: the meaning of its Being, whether it is present in the Greek, medieval or modern sense. What has traditionally been taken as a formal requirement –

---

[122] *Ibid.*, p. 685, S. 50; italics as in later edition, S. 98.

[123] Cf. later edition, S. 98.

[124] *Ibid.*, p. 685, S. 50.

when considered at all – in the sense that the presence in an opening which is necessarily common to all beings was never thought to have any content, now treats the being which it makes possible as its own formal requirement, to which it lends the content. This move is part of Heidegger's reversal of perspective, from a concern with beings to an emphasis on Being as prior.

From this new perspective, the creating and preserving of the work of art are reinterpreted in terms of the ambiguity of the characterization of the essence of art as the passive/active setting-itself-into-work *of* truth. Heidegger understands this characterization as follows: "On the one hand, it says art is the establishment of self-arranging truth in a *Gestalt*. This happens in creation as the bringing forth of the unconcealment of beings. But setting-itself-into-work also means bringing the workhood into motion. This happens as preservation. Therefore, art is the creative preservation of truth in the work. Thus art is a becoming and happening of truth."[125] The creation of unconcealment is by no means arbitrary; it merely gives to a pre-given character of unconcealment a concrete form and specifies in detail this one instance of unconcealment.

The general character of the unconcealment, the issues that are really at stake, is predetermined and merely brought out – not into the light, but *as* the light. In bringing truth out in a concretely structured form, the work comes into action as a work and preserves the truth. This working of the work is, however, the working of truth coming into work in the work. It is truth which sets itself into work in the work, at least as much as it is set into the work by the work or by the work's human creator or audience.[126]

The primacy of truth as already determined in content before the dialectic of truth and work is as clear in Heidegger's formulation, "poetry is the essence of art," as in his other formulations in terms of origin and setting-into-work. *Poetry, as naming, brings to expression the Being of beings; it does not participate in the creation of their Being. Poetry provides language in which Being can articulate itself, it explores this language and it enunciates it, but the poetic work does not participate in the epochal changes in Being as such* (the thrower in Heidegger's metaphor). "Such saying is a projection of clearing, in which it is said what the being will come into the open as. Projection is the result of a throw, which is the way in which unconcealment sends itself into beings as such."[127] The work's poetic thrust merely carries out the cast of the die in which unconcealment delivers *itself* into the being. The chips are placed long after the cast has been,

---

[125] *Ibid.*, p. 693, S. 59.

[126] Cf. later edition, S. 100.

[127] *Ibid.*, p. 695, S. 61.

unknown to the players, determined. How the being will appear in the opening is already decided by a cast which is not part of the dialectical game played by truth and the work, but which rather preconditions it.

Art, then, is not the creation of truth, not the origin and inspiration of history, but rather – as the concrete setting-into-work and poetic naming of that which is primordial – the means of executing a pre-given destiny.

> Whenever art happens, that is, when there is a beginning, a thrust emerges in history. History begins anew. History does not here mean a sequence of events, however important, in time. History is the transporting of a people into its appointed task as entrance into what has been given along with that people.[128]

Art is one of the several ways in which truth, proposed for "an historical folk," is consummated in marriage with beings. Other historical media are the action which founds a state (politics), the nearness to the "being-est" of beings (God), the essential sacrifice (Christ), thought's questioning which names Being in its question-ableness (Heidegger's philosophy).[129] Whence art receives its historical mission and why the work of art plays a special role in Heidegger's conception of history are questions which will be considered in the following chapters in terms of other Heideggerian analyses. Heidegger's closing verse taken from Hölderlin may serve here as a transition from art as an origin to the art work as place-holder.

> Reluctantly departs
> That which dwells near the origin, from the place.[130]

---

[128] *Ibid.*, p. 697, S. 64.

[129] Cf. *ibid.*, p. 685, S. 50.

[130] *Ibid.*, p. 699, S. 65.

Chapter VIII: The Thing

Thing and Stock

Heidegger's recent essay on sculpture, *Space and Art* (1969), a kind of third epilogue to the lecture we have been analyzing, provides on its concluding page a concise statement of the role of art,

> Sculpture: an embodying bringing-into-work of places and with this an opening up of realms of possible living for men, of possible persisting for the things which surround and concern men.

> Sculpture: the embodiment of the truth of Being in its place-instituting work.[131]

Sculpture creates space in the forms which comprise a piece of sculpture as well as in the surrounded forms and the encompassing region. This creating is a meaningful structuring which has its meaning in terms of a world of people and things, the very world which finds its place to exist in the space created by this sculpture. The dialectic at work here is, however, suppressed by Heidegger in this essay, so that the first sentence quoted above already prepares the way for the ontological pronouncement which immediately follows. The work of art merely embodies an externally given truth of Being by providing a place for it.

---

[131] Martin Heidegger, *Raum und Kunst* (St. Gallen: Erker Verlag), 1969), S. 13.

In the transition from the one sentence to the other, a reversal of thought takes place which robs even the initial stage of its fullness – the reversal which necessitated the preceding chapter's reinterpretation. The fullness of content which makes each piece of sculpture what it, as a unique work, is, is abstracted away until just the universal form of one of its functions remains: the creation of place. The emptiness of this characterization is then filled by Being. The original fullness of content does not, however, go unnoticed by Heidegger elsewhere; long descriptive passages repeatedly sing its praise in romantic tones. This contrast restates the paradox which the present chapter must unravel.

Throughout his writings, Heidegger takes up from different perspectives the dialectic between the opening of space, the living of man and the enduring of things. The old bridge in Heidelberg, the church bells on Christmas morn, the forest path of his childhood, the jug wet with Rhine wine and the sculptures of Giaocometti provide the material models which complement the linguistic ones of poetry and philosophy. The work of art is *creatio par excellence*: both traditionally as human creation for its own sake and ontologically as the creation of a place, be it for a simultaneously created world or for the embodiment of a given being.

The creation of space which takes this task as a *conscious* goal is, however, no longer fine art but architecture, building; where the two coincide – in the Greek temple, for instance – this is only possible because the practical goal-orientation is not total. Building, too, in its own way, provides a place for a world. A house is such a building and so is a bridge.

> The bridge is clearly a special kind of thing (*Ding*); for it gathers (*versammelt*) the Four (*Geviert*) in *such* a way that it permits it a position. But only what is *itself* a *place* (*Ort*) can encompass a position. . . . A space is something encompassed, released within a boundary, the Greek *peras*. The boundary is not that at which something ends, but, as the Greeks recognized, the boundary is that from which something begins its essence. . . . Something encompassed is always permitted and unified, that is, gathered by a place, that is, by a thing similar to the bridge. . . . The bridge is a place. As such a thing, it permits a space, in which earth and heaven, the holy and the mortal are let in.[132]

Here the Four, the interplay of heaven and earth, the holy and the mortal, is Heidegger's notion of what we have till now called the world. The bridge, as a building-thing, is place, which, as such, provides a space for the *Geviert*.

Expanding upon one of Heidegger's examples may illustrate the process at work here. The interplay of the Four cannot take place without a *place* that provides a space. But the bridge does not magically create an indistinct, homogeneous space, as if there were no difference between Heidelberg's baroque cobble and sandstone bridge and a modern steel and concrete suspension bridge, between a Black Forest hut and a Manhattan apartment. The building structures the space it presents by framing, not just spatially; by relating, not just geometrically.

Joining the shore of hillside estates and contemplative *Philosophenweg* with the shopping district, university, city hall and church, Heidelberg's historic bridge unites the private and public realms of its users, just as it originally allowed them to develop this division. In crossing the bridge, one is suspended between the eternal flow of the Neckar – wending its way between dams and locks – and the placid sky – rarely torn by military jets; between the historic-looking castle and the tourist boats. The bridge does not physically encompass town and country, castle and barge, but in providing a place for them to converge and appear, it lets them interrelate and become what they are.

The sculptural work, as a place, creates space in which a world can come to be. A building approaches this task systematically and designates by its own

---

[132] Martin Heidegger, "Bauen Wohnen Denken" (1951) in *Vorträge und Aufzätze*, Bd II (Tübingen: Neske, 1954), S. 28f.

structure the outlines of the world which can occupy its room. These are two examples out of many of things which create space and structure a world within it. A jug is another clear case of something which not only defines a place, but in so doing, creates a space: not just the volume of its hollow, but the world in which it relates to heaven and earth as producers of its substance and contents or to the holy and mortal to whom it grants offerings and nourishment.

Heidegger finds the unity of the jug gathered together in its ability to grant the gift of its contents. In gathering itself together, the jug gathers together the world, the Four:

> That which is gathered in the gift gathers itself together by appropriatingly letting the Four abide. This manifold simple gathering is the jug's essence. The German language has an old word for what gathering is. The word is *"thing."* The essence of the jug is the pure bestowing gathering of the singular Four in an abiding. The jug is essentially present as thing. The jug is the jug as a thing. But how is the thing essentially present? The thing things. The thinging gathers. it gathers, appropriating the Four (*das Geviert ereignend*), the Four's abiding in something momentary: in this or that thing (*Ding*).[133]

To be a jug means to let a world come to be, which then lingers as a unity in the unity of the jug. The jug, like a work of art, but in its own way, is creator and preserver of a world and is this essentially.

The phrase, "appropriating the Four," indicates that appropriating the Four is part of the jug's function as a "thing." The term "thing" does not refer to everything, but only to those things which "thing" in the verbal sense that they gather the Four's abiding in themselves in letting the Four come to abide, in "appropriating" (*ereignen*) it. Still, this covers a whole host of "things" according to Heidegger, including the jug and the bench, gangway and plough, tree and pond, creek and mountain, heron and roe, horse and steer, mirror and banner, book and picture, crown and cross.[134] The last four of these we have already met in their more allegorical expressions.[135]

The analysis of the jug led to a notion of thing which, however, is not applicable to all things, not even always to jugs. The general term, "thing,"

---

[133] Martin Heidegger, "Das Ding" (1950), *ibid.*, S. 46.

[134] *Ibid.*, S. 55.

[135] Cf. "The origin of the work of art," *op. cit.*, S. 50.

has changed its meaning with the historical changes in the meaning of its synonym, "being." Heidegger's usage is emphatic, utopian. The jug is not a thing in the Roman sense of *res*, nor the way the Middle Ages conceived of *ens*, nor even in the modern conception of object. The jug is a thing insofar as it things. From the thinging of the thing the presence of what is present in the manner of a jug first appropriates itself and determines itself.[136]

The tendency of our times militates against things being present as thinging things. Heidegger's essay on the thing is, in fact, introduced by a consideration of our times in terms of the collapsing of all distances, the very opposite of the nearing which is essential to thinging:

> The fact that the recent increasing elimination of separation has not resulted in nearness has brought distancelessness into dominance. With the lack of nearness, the thing remains annihilated in the aforesaid sense of thing. But when and how *are* things as things? We ask this in the midst of the domination of distancelessness. . . . What becomes thing appropriates itself out of the Ringing (*Gering*) of the mirroring-play (*Spiegel-Spiels*) of the world. Only if, perhaps sometime, the world worlds as world, then will the ring shine, which rings around the Ringing of earth and heaven, the holy and the mortal in the ring of their simplicity.[137]

This is the historical factor, which, when introduced into the analysis of the work of art, seemed to necessitate a reinterpretation in terms of a Being given independently of the work. What determines historically whether the things in an age will be things in the Roman, medieval, modern or emphatic sense?

This determination must, it would seem, be prior to the existence of the thing – in whatever sense – and therefore independent of it. Heidegger indicates that the thinging of the thing depends upon the worlding of the world, a worlding of which our world does not seem to partake. This answer, if accepted, pushes the historical question further in two directions:

1. Is the worlding of the world independent of the thinging of the things in it?

2. What determines historically whether the world worlds or not?

---

[136] *Ibid.*

[137] *Ibid.*, S. 54f.

Technological Being

Is the worlding of the world independent of the thinging of the thing? Penetrating the thickets of Heidegger's terminological jungle from the thing itself to seeming tautologies, one returns to the starting point to discover that one has traced a path from the thinging of the thing to the worlding of the world without ever leaving one's seat. Heidegger's bombardment of terms, distortions of the already flexible German syntax, may not reveal an experiential wealth behind each word in the concise presentation of the last few minutes of his lecture on the thing, but one can, at least, reconstruct a link between what he calls thing and world.

We have already seen that the thing, in being a thing, in thinging, grants a permanence to the Four in its unity. "Thinging, the thing abides the unifying four, earth and heaven, the holy and the mortal, in the simplicity of its self-unifying Four."[138] This unity is a dynamic one, in which each of the four constituents appropriates itself within the unity of the Four. "Earth and heaven, the holy and the mortal belong together, unified with each other from themselves and on the basis of the simplicity of the unifying Four. Each of the four in its own way reflects itself appropriately within the simplicity of the Four. This mirroring is no presentation of a representation. The mirroring appropriates, while clearing each of the four, each's own essence to one another in the simplistic unification."[139] This mirroring process can be taken to be the experiential core of what Heidegger is struggling to communicate – not just on these pages, but in all the descriptions and schemes of conceptualizations he has published.

In the term *mirror* one glimpses an unfocused image of Hegel's speculative thought, which once objectified itself with the help of a possibly less metaphorical term, *mediation*. What the newer jargon is meant to express must be experienced by each reader in the models referred to above: sculpture, the bridge, a jug, the church bells, a path through the fields. The play of mirroring, conceived on the grand scale, the appropriating mirroring-play of the simplicity of earth and heaven, the holy and the mortal, is called the world.[140]

This provides the answer to the question of the dependence of World on thing. The worlding of the world is practically nothing but the thinging of the thing disguised in different nomenclature. The synonymity is not exact;

---

[138] *Ibid.*, S. 50.

[139] *Ibid.*, S. 52.

[140] *Ibid.*

subtleties provide distinctions. The thing is a place which provides Space and provides permanence for the world, but it is more than a creative preserver in this sense: it fulfills this formal function, but also provides a content and does this precisely by bringing the four of the Four into a mutually mirroring play of unity. This play is the world. Translating again into Hegelian language, the thing is a moment in the dialectical (mirroring) process (play) of the world's self-constitution (worlding).

The remark which raised the question of the relationship of world and thing was, however, somewhat more complex. It refers to the ring and the Ringing: "What the thing becomes appropriates itself out of the Ringing of the mirroring-play of the world." The new terms are introduced in another series of defining pseudo-equations in which the processes of objects which are nothing but their processes are given new names, to which the same trick is played over and over again!

> The fouring is present (*west*) as the appropriating mirroring-play of the simply inter-reliant. The fouring is present as the worlding of world. The mirroring-play of world is the propriating dance of Appropriation (*Reigen des Ereignens*). Thus the propriating does not primarily encircle the Four like a hoop. The propriating is the ring that rings by playing as the mirror. . . . The collected essence of the thus ringing mirroring-play of the world is the Ringing.[141]

What was concluded about the relationship of thing to world can be repeated for that of world to Ring. Through algebraic cancellation of the middle terms in these relationships, we see that the Ring is nothing transcendentally independent of the thing.

The determination of thing by Ring is misunderstood without its opposite, the characterization of the Ring as an explication of the thing. On the one hand, the thinging of the thing appropriates itself out of the mirroring-play of the Ring of the ringing. On the other hand, the thing lets the Ring abide in something momentary from the world's simplicity.[142] The two poles of the dialectic can be held apart if we say that a world finds its place and permanence in a plurality of things, that the simplistic isomorphic image must be corrected by a many-to-one relationship of things to world. The world in which, for instance, the residents of Heidelberg are united with their Neckar Valley surroundings, with their heavens and the holy, finds its space in many places, its permanence in many things: bridge and alleyways, church and

---

[141] *Ibid.*, S. 53.

[142] *Ibid.*

houses, castle and university, the sculptures of the gateways and the lyrics ringing through the beer halls. The Ring encircles all this in the aura which is old Heidelberg, according to the dialectic of whole and part which Hegel once outlined in that city's lecture halls and journals.

But the relationship of thing to Ring as its explication is capable of a less dialectical interpretation as well. Perhaps Heidegger feels that the Hegelian method of analysis distorts by imposing an external scheme. An explication of something that remains with the thing itself, with its own immanent categories, can be understood as a phenomenological search for essence, a return "to the things themselves." How does one articulate the essence (*Wesen*) of world without concepts heterogeneous to the phenomena? Heidegger's attempt: "World is present (*west*) by worlding. This means, the worlding of the world can neither be explained nor founded on the basis of anything else."[143] By articulating the world's form of presence with the use of the verbal form of its name, no new determinations are imposed. Similarly, Heidegger chooses the same technique in uncovering the essence of the thing on its own terms. He defends this approach as follows: "If we let the thing be present in its thinging out of the worlding world, then we think about the thing as the thing. . . . If we think about the thing as thing, then we protect the essence of the thing within that region from which it is present."[144]

Clearly both tendencies – phenomenological and, where that proves inadequate, dialectical – are present throughout Heidegger's analyses, as any situating of his thought in the history of philosophy immediately reveals.[145]

Whether and how the two interpretations are compatible is perhaps less urgent for us than whether either satisfactorily responds to our original problem, the historical ontological question. We asked if the worlding of the world was independent of the thinging of the thing in the hopes that an affirmative answer would yield a clue to the central difficulty: what determines historically whether the world worlds or not? The doubly negative answer, that neither thing nor world determines the other, that they are inextricably mutually dependent within the dialectic which they both are, leaves us without a clue,

The paradox is this: on the one hand, a jug is a thing insofar as it things; for a thing to thing, the world must world; but the world does not world – that is the problem of our times. On the other hand, the presence of what is present

---

[143] *Ibid.*, S. 52.

[144] *Ibid.*, S. 53.

[145] Cf. *Zur Sache des Denkens, op. cit.*, p. 64, S. 70f for Heidegger's own statement of this.

– the Being of beings, whether a jug is a *res*, an *ens*, an object or a thinging thing – is determined, according to Heidegger, by the thinging of the thing and the thing in turn by the Ring of the mirroring-playing of the world. The historical problematic remains unclarified: what determines when the world will world? To this we can add the question: What is special about thinging and worlding as modes of the Being of things and worlds such that these modes determine all other modes even when only a different mode is present? What model will help us to unwrap this paradox? Where does Heidegger find a hint?

Forgetfulness of Being

The art of church bell ringing provides Heidegger with a striking example of a place for the Four to come together and receive a structuring:

> The mysterious unity in which the church holidays, the festivals, the passing of the seasons and the morning, afternoon and evening hours of every day are merged into one another so that always *one* single tone rings through the young hearts, dreams, prayers and games – it is surely this which conceals itself with one of the most magical, holiest and most lasting secrets of the steeple, in order to bestow it, ever transformed and unrepeatable, until the last chime in the highlands of Being.[146]

The bells which ring out across the village before dawn Christmas morning are a highly structured, artistically unified combination of the tones which strike the hours of the day and ring in the seasonal holidays throughout the year. Each tone has its meaning in the rhythm of the community, in the life of man. Tradition has blended the various clangs into a harmony which plucks the strings of melancholy in the heart of a man who has been immersed in two world wars since swinging carefree from the bell tower ropes as a child. The sounds which once accompanied the experience of awe, the play of children, the excitement of Christmas, the burial of victims of war – soldiers, friends – the routines of a bygone life, these chimes now bring back into existence the life and world which they long ago accentuated, structured.

---

[146] Martin Heidegger, "Vom Geheimnis des Glockenturms" (1956) in *Martin Heidegger zum 80. Geburtstag* (Frankfurt: Klostermann, 1969) S. 10.

The familiar path through the fields, from home town to its neighbor, plays a similar role in providing a place for the meeting of man and nature: heaven and earth, the holy and the mortal. Out of the overtones and the over-all tone which the path lends to whatever takes place on it, one can read the nature of those constituents of the Four. The path can be made to speak by a kind of reversed dialectic, a reconstruction. Returning to town at night, retracing his footsteps of earlier and much earlier, Heidegger hears the church bells and listens to the field path:

> Slowly, almost hesitantly, eleven hour strokes echo in the night. The ancient bell, on whose chord the tender hands of choir boys had often been rasped, trembles under the poundings of the clapper, whose ominous, comical expression none could forget. . . .

> The exhortation of the path through the fields is now quite clear. Does the soul speak? Does the world speak? Does God speak? Everything speaks of the refusal of the sight of one and the same thing. The refusal of sight does not take. The refusal of sight gives.[147]

The serious thinker of Being and the playful choir boy he once was, the harmony of a romanticized farm life and the alienation of technological existence, the holy ringing of the bells and the death of God: the contrasts repeat the message of insight refused over and again. This is the message Heidegger hears everywhere and tries to unravel. The obfuscations of the good life, the just society, the presence of God – all point to their determinate negations by concealing them; the only hint of Being we have is our forgetfulness of it, as *SuZ* points out on its first page. The forgetfulness, expression of a refusal, is the very measure of good and evil, of man and world, as He may once have been.

> The measure consists in the way the God who remains unknown is manifested by heaven *as* this unknown. The appearance of God by heaven consists in a covering up, which allows to be seen that which conceals itself, but does not allow it to be seen by trying to tear the concealed out of its concealment, but only allows it to be seen by protecting the concealed in its self-concealing. Thus shines the unknown God as the unknown by the manifestness of heaven. This appearance is the standard by which man measures himself.[148]

---

[147] Martin Heidegger, "Der Feldweg" (1949), *ibid.*, S. 14f.

Leaving for now the things which reveal forgotten, refusing, self-concealing Being, we will see in the following chapter how Heidegger attempts to break through the appearances of our times to analyze Being-itself as the refusal which gives us our non-worlding world. Perhaps the reversal of perspective will help clarify the fundamental ambiguity in our Heidegger interpretation. We have, that is, repeatedly seen that beings play a role in the creation of Being. At first it seemed as if Being were merely a characteristic of beings as they exist and interact with one another, dialectically creating their own world. But then Heidegger threw in cryptic remarks pointing to a somehow pre-given Being which determines beings and their worlds. The difficulty of interpreting these relationships seemed almost to grow rather than disappear, calling for a new approach, a look at the over-all structure of Heidegger's ontology in its most mature expression.

---

[148] Martin Heidegger, "Dichterisch Wohnet der Mensch" (1951) in *Vorträge und Aufzätze*, Bd II (Tübingen: Neske, 1954), S. 71.

Chapter IX: Being-Itself

The History of Being

That Heidegger's is a negative ontology can be seen throughout the development of his thought and on various levels. His destruction of the history of philosophy and his recurrent criticism of prevailing ontological notions represent only the most superficial aspects of this negativity. Already there between the lines in *SuZ*, negativity is thematized in *What Is Metaphysics?* and repeatedly reflected upon thereafter. The central ontological categories in all Heidegger's writings contain their negation as a largely unexpressed, yet fundamental, correlate: *Sinn von Sein* and *Seinsvergessenheit*, *Unverborgenheit* and *Verbergung*, *Sein* and *Nichts*, *Sicht* and *Versicht*, *Ereignis* and *Enteignis* (the meaning of being and the forgetfulness of Being, unconcealment and concealing, Being and nothingness, sight and refusal of sight, the appropriation and the expropriation). The structure of this negative ontology is probably most explicit in the minutes to a seminar in which Heidegger openly reflected upon his own path of thought, but the comments there rely upon a familiarity with his writings.

SuZ is an attempt at a negative ontology in the sense that it tries to develop an ontology under the condition that ontology is today impossible. The experience which underlies *SuZ*'s attempt to work out the question of the meaning of Being is that of the forgetfulness of Being. This oblivion of Being is, however, not conceived as the result of laziness or neglect on the part of philosophers,[149] but as an essential consequence of the nature of Being itself. If God appeared in the Middle Ages in His creations, He is now only there in the form of a refusal to be seen; if Being appeared to the pre-Socratics in

---

[149] The forgetfulness of Being is, however, observable in the situation of academic philosophy. The neo-Kantian bracketing of ontology, which led Husserl away from Being, and the contemporary rejection by positivism of what it considers metaphysical, resulting in a deep-seated scorn of Heidegger's work, are perhaps extreme examples. Cf. *On Time and Being, op. cit.*, p. 44, S. 47.

beings, it is concealed to us by them. The sight (*Sicht*) which is the appropriate access to Being[150] is now determined by the *Versicht*, in which alone Being is today present as absent, as in oblivion. Heidegger's methodology takes this circumstance into account from the start.

Calling the method of his ontology "phenomenology," he characterizes the phenomena which he seeks as follows:

> Within the horizon of the Kantian problematic an illustration of what is conceived phenomenologically as a 'phenomenon', with reservations as to other differences, can be given if we say that that which already shows itself in the appearance as prior to the 'phenomenon' as ordinarily understood and as accompanying it in every case, can, even though it thus shows itself unthematically, be brought thematically to show itself; and what thus shows itself in itself ('forms of intuition') are phenomena of phenomenology.[151]

The phenomena of phenomenology ordinarily present themselves by implication alone, perhaps concealing themselves in the very act of helping ordinary "phenomena" come to thematic appearance. This is a purely formal concept of phenomenon, specifying neither which being – or Being – is referred to, nor the sense in which it "always prior and accompanying, but unthematically" shows itself. The concept is concretized in various ways in Heidegger's writings and it may be helpful to glance at these before turning to the lecture on *Time and Being* and the related seminar.

Perhaps the main lesson for ontology in *SuZ* itself is that Being is not something divorced from beings, but rather the structure of their own presence. The wholeness of *Dasein*, for instance, is sought in a series of structural analyses, which characterize its Being-there as Being-in-the-world, care, temporality. Heidegger's subsequent analyses of the work of art, the jug, etc. never leave the beings under analysis to make deductions from a higher being of some kind. The ontological structures are existentials, structures of the form of Being of the beings, explications of what was always non-thematically inherent.

> 'Behind' the phenomena of phenomenology there is essentially nothing else; on the other hand, what is to become a phenomenon can very well remain concealed. . . . Our investigation itself will show that the meaning of

---

[150] Cf. *Being and Time, op. cit.*, p. 187, S. 147.

[151] *Ibid.*, p. 187, S. 147.

phenomenological description as a method lies in *explication*. The *logos* of the phenomenology of *Dasein* has the character of a *hermeneuein*, through which the authentic meaning of Being and the fundamental structures of his own Being are made known to *Dasein*'s understanding of Being.[152]

That this emphasis on explication of the things themselves is still important in Heidegger's later writings was seen in the preceding chapter. After considering alternative interpretations of the essay *The Thing*, we concluded that the Four, mirroring-play, world and appropriation stand in a similar relationship to the thing as care, etc. do to *Dasein* or world and earth, creating and preserving do to the work of art, namely as names for moments of its processual Being.[153] In each case, it is a matter of explicating the form of unconcealment of something which is, particularly in our era, concealed.

Heidegger's publications shortly after *SuZ* develop themes of negativity only touched upon in the context of the *Dasein* analysis. *On the Essence of Truth* elaborates the notion of ontological truth as unconcealment in its interplay with concealment; this notion we have already met in connection with the work of art as the setting-into-work of truth. *What Is Metaphysics?* attacks the central problem of unconcealment from another angle: reversing *SuZ*'s perspective on the phenomenon of *Angst*. In the earlier presentation, *Angst* was important because it revealed the world as world (in the sense of world used in *SuZ*) to *Dasein*[154] by negating his everyday involvement with individual beings. The negation, which is non-thematically at work in the revelation of *Dasein*'s context of meanings, is, as such, thematized in *What Is Metaphysics?*:

> In *Angst* there is a retreat from. . , though it is not so much a flight as a spell-bound peace. This retreat from. . , has its source in Nothing. The latter does not attract; its nature is to repel. This repelling from itself is as such the relegating reference to the vanishing beings in totality. This relegating reference to the vanishing beings in totality, as which Nothingness crowds round *Dasein* in *Angst*, is the essence of Nothingness: Nihilation.[155]

---

[152] Thus, although SuZ's "fundamental ontology" was more a preliminary analysis which had to be repeated later from an entirely different approach, the character of phenomenological description remains explication. *Ibid.*, p. 60f, S. 36f.

[153] Cf. e.g., *On Time and Being*, p. 32, S. 34.

[154] Cf. *Being and Time*, p. 232, S. 187.

Nihilation is the precondition for the analysis of world in *SuZ*. In referring attention to the uncomfortable world of *Angst* in which all things lose their value for *Dasein*, Nihilation directs attention away from itself. Necessarily accompanying the revelation of the world of *Angst*, Nihilation is itself self-concealing. The two processes, revealing and concealing, are really one motion. This is the characteristic finitude of revelation: no revealing without concealing, it is a theme of Heidegger's which recurs under many titles: *Wahrheit* and *Irre*, *Geben* and *Entzug*, *Lichtung* and *Bergung* (truth and error, granting and removing, clearing and concealing). Nothingness is here still a "phenomenon" as defined in *SuZ*, but it is now explicitly a structure of Being itself, of unconcealment, no longer specifically of *Dasein*. In Heidegger's mature terminology, Nothingness is a giving which gives the revelation of the world as its anonymous gift while holding itself back. The lecture, *Time and Being*, tries to thematize the way in which Time and Being are anonymously given in this way, or in which they give themselves. This is, of course, the problematic we have already met. In his essay on art, Heidegger explored the work of art as one way in which truth – and that means a world, Being, and an historical epoch, Time – is given. Later, he generalized the notion of an art work as dynamic unity of world and earth to his utopian conception of a thing, which gives itself in the playful mirroring of the Four.

The interpretational ambiguities which made Heidegger's thought so inaccessible or so misunderstood stem largely from the negative form of his ontology, which paradoxically yearns to negate *to on* (the being) instead of classifying it. *Dasein* is not really, authentically, that which it primarily and for the most part of necessity is; what a thing is, is determined by the worlding of a world which unfortunately does not world; and what *is* all around the earth today has progressively obscured itself for more than 2000 years by the way in which everything "is" in a less emphatic sense. During the development of Heidegger's thought, the negative aspect gains gradually in historical sharpness. If *SuZ*'s explications necessitated the overcoming of a rather ahistorical inauthenticity, the work of art was already an historical deed, overturning systems of values and revamping the world-views of historical folk. Finally, the notion of thing embodies a negation of our technological era, represented in everything from television to atomic bombs. But the increased emphasis on history does not clarify Heidegger's meaning; history and Being merely become inextricably entangled in the paradoxes, adding to the confusion.

---

[155] Martin Heidegger, "What is metaphysics?" in *Existence and Being* (Chicago: Gateway, 1965), p. 338; Cf. Martin Heidegger, "Was ist Metaphysik?" in *Wegmarken* (Frankfurt: Klostermann, 1967), S. 11.

The keystone to Heidegger's paradoxical negative ontology may be seen in the notion of the end of the history of Being, a phrase that literally combines negativity, history and ontology. The structure of the negative ontology was clearly presented in the 1949 lecture series, "Glimpse into that which is": *The Thing, The Gestell, The Danger* and *The Reversal*, but is more thoroughly discussed in *Time and Being* (1962) and the minutes to the accompanying seminar. In the earlier lecture series, the utopian conception of the thing was first presented; then the nature of technology was analyzed as the *Gestell*, presenting a contrast between the priority of the thing in the worlding of the world and its total subjugation to the subject in the technological context that determines our present world. That the thing cannot be what it ideally is, that Being is hidden by the technological essence, represents the danger (*Gefahr*) of our times and makes the task of thought the return (*Kehre*) into Appropriation (*Ereignis*). From the start, there are perplexing problems to this scheme, already apparent even in so brief a summary: What is Appropriation and what can it mean to return into it? If the *Gestell* is a form of Being, how can it be said to hide Being and what right has Heidegger to reject the technological form of beings and Being in favor of an "authentic" form, which often seems merely romantically utopian? These and similar questions lead us into the problematic of the history of Being.

The lecture, *Time and Being*, is composed of three sections: an introduction, a presentation of the problem-complex named in the title, and the analysis itself, which moves from Being and Time to Appropriation. The introduction hints at the underlying problem as well as at methodological considerations. The task of philosophy is, it suggests, conditioned by the contemporary scene, from which it receives its necessity both in the sense of motivation and of restriction:

> It might be that this kind of thinking is today placed in a position which demands of it reflections that are far removed from any useful, practical wisdom. It might be that a kind of thinking has become necessary which must give thought to such matters from which even painting and poetry and mathematical physics receive their determination. Here too we should have to abandon any claim to immediate intelligibility.[156]

The task of thought is still to help live the reflected life, to be wise in the ways of the world. But this task cannot be accomplished directly. It requires an analysis of that which *determines* contemporary appearances – such as

---

[156] "Time and Being" in *On Time and Being*, p.1; "Zeit und Sein" in *Zur Sache des Denkens*, S. 1.

Klee's paintings, Trakl's verses, Heisenberg's scientific theories – and *conceals* itself in them. This is the question of Being in an historical and negative formulation. The formulation implies for Heidegger that the analysis must be one of "Being without reference to a grounding of Being in beings" and that the lecture which presents this analysis cannot be immediately understood as a string of propositions, but only through the mediation of the *experience* of thinking about Being.

The question of Being is historically situated because Being is itself historical. Firstly, as Heidegger claims immediately after the introduction, Being means presence and is thereby related to Time as the unity of the present with the past (presence refused) and the future (presence withheld). Secondly, Being as presence has taken many historical forms, as can be seen in the history of philosophy: Plato's *ousia*, Aristotle's *energia*, Kant's *Gegenständlichkeit*, Hegel's *Gesetztheit* and our technological era's calculable material. Heidegger must therefore explicitly include the historical question, whence the unity of all appearances in a given age, with his question of Being. This he does in his late lecture by asking not "What is Being?" but 'How is Being given?" (*Wie gibt es Sein?*) The appearances characteristic of our world – mass media, the nuclear threat, totalitarianism, subjectivism, the crisis of the cities – share an essential rootedness in the contemporary form of Being: presence as pliable raw material for the uses of people. The question – both historical and ontological – then is: how does it come to be that presence is *now* given in this form? This question forms the starting point for the main text of the lecture in terms of the relationship of Time and Being – history and presence – and the phrase, "being is given."

In *SuZ* the historical character of the question of being paralleled Husserl's conception of an "archaeology of meaning."[157] To *Dasein*'s inauthentic fallenness in the world of objects corresponds his unquestioning acceptance of a tradition, a set of prejudices, including the prevailing philosophical tradition with its implicit ontology. The amalgam of categories, beliefs and outlooks that have been accumulated in the largely unreflected transmission of wisdom from one generation to the next obscures through its own dialectic both its real value and its problematic character:

> The tradition which comes to dominate makes what it
> 'transmits' so inaccessible proximally and for the most part
> that it tends rather to conceal. Tradition takes what has
> come down to us and makes it commonsensical, blocking

---

[157] Cf. Edmund Husserl, "Der Ursprung der Geometrie" in *Die Krisis der europäischen Wissenschaften und die Transzendentale Phänomenologie* (the Hague: Martinus Nijhoff, 1954).

our access to those primordial 'origins' from which the categories and concepts handed down to us have been in part quite genuinely drawn. Indeed, it makes us forget that they have had such a source.[158]

The posing of the question of Being was meant to reawaken an awareness of this process. The task remaining after the analyses published in *SuZ* was to remove the obscurities, in terms of the question of Being, in the great metaphysical systems of Hegel, Kant, Descartes and Aristotle, finally returning to the original experience of Being as presence in the pre-Socratic philosophies. According to the plan of *SuZ*, this "destruction" of the history of ontology was meant to be preceded by a discussion of Time and Being which would take off from *SuZ*'s analysis of *Dasein* and temporality. This discussion was meant to give a formal answer to the question of Being which would then orient concrete textual analyses of the history of ontology, allowing them to uncover forgotten potentialities and thereby preparing the way for a new relationship to Being.[159] The answer to the question of Being given in the section "Time and Being" was thus envisioned as neither radically new nor as a final panacea for ontology and society.

Of course Heidegger never carried out this plan in the original order. After publishing *SuZ* ("Part I") with its analyses of *Dasein* and temporality, he suppressed the section on Time and Being, which was to provide the formal answer to the question of Being, and for the next 42 years (1927 - 1969) published thousands of pages of historical critique before printing the lecture to which we next turn. At the end of our discussion of this lecture, we will have to deal with the consequences of this reversal in publication schedule for the formality of Heidegger's final answer to the question of Being.

Time and Being begins the body of its analysis with the insight into the forgetfulness of Being in general and the historical process of progressively covering the understanding of Being in the thoughtless acceptance of new ontological categories at face value, the confusion of mixing incompatible metaphysical systems and the distortion inherent in translating terms (and the thought they embody) from Greek to Latin to German, French and English. Granted that *Being* is historically *given* as presence – and the lecture's linguistic motor force is: "*Es gibt Sein*" – then, it seems, the process in which Being is so given (*das Geben*) is unknown. Being is an anonymous gift that we receive

---

[158] *Being and Time*, p. 43, S. 21.

[159] This relationship of formal theory to the texts which form the theory's pre-history is not unlike that of *Capital*'s first volume (the theory of value and of capitalist production) to its projected fourth volume (the notes of which form three volumes of critique: *Theories of Surplus Value*).

without being aware of the giver or his giving, for we are exclusively involved with the presence of beings.[160] That which is sent the way Being is sent, without the sender or the sending appearing, is called *Geschichte*, history, from which Heidegger derives his notion of *Seinsgeschichte*, the history of Being.

Heidegger introduces his notion as follows:

> The history of Being means destiny of Being in whose sendings both the sending and the It which sends hold back with the announcing of themselves. To hold back is, in Greek, *epoche*. Hence we speak of the epochs of Being.... The epochs cover over each other in their sequence so that the original sending of Being as presence is more and more obscured in different ways. Only the gradual removal of these coverings – that is what is meant by the "destruction" – procures for thinking a preliminary glimpse into that which reveals itself as the destiny of Being.[161]

Here themes from *SuZ* are repeated: the word play with *Schicken, Geschick, Geschichte* and the "destruction" of the process of covering up. However, the priority has been reversed. The historical destruction now only gives a preliminary glimpse, summarized in the final sentence of the excerpt. The real task is to uncover what is still hidden even when the various stages of the history of ontology have been explicated and it is clear how Plato conceived of Being as *idea*, Aristotle as *energia*, etc. and it is also clear how these are all related forms of presence. Still hidden after the destruction are the giver and the giving, which give Being in its various historical forms of presence.

We have already taken a look at one of Heidegger's attempts to work out the structure of the giver and giving in which Being is given. In his lecture on the thing, Heidegger described the Being of a thing, a jug, as its "thinging": "What becomes a thing, appropriates itself out of the Ringing of the mirroring-play of the world." This Being is given in the worlding of the world, which in turn was attributed to Appropriation: "The mirroring-play of world is the propriating of Appropriation." Viewed within the "*Es gibt Sein*" formula, the Appropriation is the giver, which gives the jug its Being as a thinging thing, and the giving takes place in the mirroring-play of the worlding of the unity of heaven and earth, the holy and the mortal. The case of the jug as thing was, however, distinguished as "utopian" from the normal case in which the jug is an object of utility waiting to be fit into someone's plans.

---

[160] "Time and Being," p. 9, S. 8.

[161] *Ibid.*, p. 9, S. 9.

Before following Heidegger's argument from Being and Time to Appropriation (in the next section), it will be useful to outline the role Appropriation must play in Heidegger's system. This role is intimately involved with the contrast between the contemporary and utopian forms of Being. In fact, the contrast has methodological implications for the working out of the structure of Appropriation within the context of the history of Being. Although the contrast was foreshadowed in the conceptual pair, authentic/inauthentic, in *SuZ*, the analysis of the work of art provided the utopian notion that acted as a catalyst to Heidegger's progress after 1935. The importance of this can be seen in Heidegger's statement that the relations and contexts that make up the essential structure of Appropriation were worked out between 1936 and 1938.[162] The thirties were certainly a time of crisis in Germany, when everyone was searching for alternatives of one kind or another on a world-historical level and Heidegger was more an extreme example of this than the exception he is often taken to be. His concept of the *Gestell* is a central, but ambiguous, key to his approach to this task.

In the 1956 "Addendum" to his essay on art, Heidegger pointed out the contrast between the work of art and technological beings in terms of his concept of the *Gestell*. The essay refers to the *Gestell* "as which the work is present insofar as it erects and produces itself." To this Heidegger comments,

> The word '*Gestell*' which is later used explicitly as the term for the essence of modern technology was used in the essay on art in this sense (not in its common meanings of book case, montage, etc.). This connection is essential because it is part of the history of Being. . . . On the one hand we must avoid the specifically modern connotations of thesis or positing (*Stellen*) and montage (*Gestell*) in *Fest-stellen* and *Ge-stell* in *The Origin of the Work of Art*. On the other hand, we must not overlook the fact that and the degree to which the Being which determines the modern age comes as *Ge-stell* out of the Western destiny.[163]

There is a certain ambiguity to Heidegger's use of the term *Gestell*: it can refer either to the utopian worlding of the world in a work of art or also to the essence of modern technology. This ambiguity is not a matter of conceptual sloppiness or coincidence, but is supposedly the result of an essential relationship within the history of Being. In commenting on the introductory remarks to *Time and Being*, Heidegger states that the *Gestell*, as the preliminary appearance (*Vorerscheinung*) of the Appropriation, is also that which

---

[162] *Ibid.*, p. 43, S. 46.

[163] Later edition of "Der Ursprung des Kunstwerkes," *op. cit.*, S. 72, S. 97f.

necessitates the attempt to bring into view that which is around the earth today.[164] The *Gestell* involves the glimpse that phenomena like the work of art give us of Appropriation, which sends the forms of Being, and it involves the latest form of Being, which we have been given in the modern technological age and which represents the danger of our times.[165]

The *Gestell* as the essence of technology is simply the latest, most extremely subjectivistic form of Being. However, in the work of art the *Gestell* is part of the worlding of the world; a stage in the process in which Appropriation sends Being, and not merely that which is sent. In this latter sense, the analysis of the *Gestell* represents a move beyond that which is revealed in the history of Being and towards an uncovering of the giving and the giver, which normally conceal themselves. The transformation of Being into Appropriation thus steps outside of the epochal transformations of Being, from which it must be distinguished.[166] This stepping outside, which Heidegger refers to as the reversal (*Kehre*) or the step back (*schritt zurück*), constitutes the leaving behind of metaphysics.

In turning away from Being as the Being of beings, Heidegger leaves the metaphysical tradition which still influenced the original plans for its own destruction in *SuZ*, and tries to think about Being in terms of Appropriation or "Being-itself," which gives Being. Heidegger's notion of Being as Appropriation is thus qualitatively different from all the previous notions of Being as *idea, energia*, etc. The analysis of Being as sent demands a consideration of the sending in which all the previous notions of Being have been sent, it could be called a meta-ontology, a reflection upon the preconditions of all systems of metaphysics. The Olympian position Heidegger attains in his meta-ontology secures him from the charge of relativism; in explaining the multitudes of ontologies, he need not compete on their level.

Heidegger defends his system with a familiar tactic. Hegel's idealistic theory was the final word because it incorporated the end of the history of the concept. Marx's materialism founded the objectivity of its class analysis in its perspective from the potential end of the history of class domination. Analogously, the thinging of the thing is not utopian in the sense of wildly romantic wishing, but, as part of Appropriation, is founded in the end of the history of Being and is thus meta-levels beyond technological beings with

[164] "Time and Being," p. 33, S. 35.

[165] *Ibid.*, p. 53, S. 57.

[166] *Ibid.*, p. 52f, S. 56.

their positivistic *fundamentum in re* rather than in the giving of the meaning of *res*.

Heidegger's end of history is, formally at least, more like Marx's than Hegel's. Where the conservative idealist thought that progress had already reached its goal, his materialist critic viewed the end of the previous form of history (unplanned "prehistory") as the occasion for a qualitatively different form: truly human progress based on a conscious response to concrete needs. The Heideggerian step into Appropriation is not viewed as a stopping of the sending of Being in new forms, but as a dawning awareness of the process of this sending, an end to the forgetfulness of Being and thus to metaphysics as the reflection of this forgetfulness in philosophy.[167] While people as mortals remain limited in their relation to Being even when thought has experienced the Appropriation – and this notion of human finitude is perhaps the crux of Heidegger's rejection of Marx – Heidegger does retain an optimism that the danger which has been growing since Plato can be abated by ending the self-concealing of Appropriation

Meta-ontology

We have seen that the Appropriation is that which gives Being in various historical forms and which conceals itself and its giving in the gift of Being. The task of thought today is thus, according to Heidegger, to "step back" from the view of Being in terms of the beings which it lets be and to turn to the process in which Being is itself given, to experience Appropriation. Keeping this in mind, we can understand Heidegger's presentation of Appropriation in *Time and Being*. The underlying experience of negative ontology is *Versicht, Vergessenheit, Verbergung*. Concealing takes place at every stage of Heidegger's argument: Being is sent to us, but the giving keeps to itself. Time is passed to us, but the present is withheld from us in the future and refused us in the past.[168]

The point of the step back is to reveal these self-concealing phenomena by understanding them in their self-concealing. The stepping is thus ambiguous, involving a *movement into* the underlying phenomena (in the phenomenological sense part of a *stepping back* to let them reveal themselves). The second tendency seems to gain in priority in Heidegger's later writings, contributing

---

[167] *Ibid.*, p. 50, S. 53f.

[168] *Ibid.*, p. 22, S. 23.

to their mystical tone because of its passive receptivity and consequent incommunicability.

To make intersubjective sense out of this two-pronged approach to the analysis of Appropriation – the retreat from metaphysics to step back into underlying Appropriation, to let it show itself as itself free from our metaphysical language – we are forced to stress the moment of phenomenological explication (*Auslegung*). The other moment, letting appropriation show itself freely, we can only understand in terms of Heidegger's frequently repeated warning not to understand the phrases "It gives Being" and "It gives Time" as propositions with subject, verb and object, but rather to understand them on the basis of that which is given – Being and Time – and how it is given – sending and passing.

Time and Being begins its struggle with the related problems of the relationship of Time and Being – history and presence – and the way in which Being is given in terms of Being as presence. The determination of Being as presence, assumed valid for all epochs in the history of Being, can be taken as a result of Heidegger's historical studies elsewhere or – like the determination of *Dasein*'s Being as Existence in *SuZ* – as a leap into a hermeneutic circle, not to be deduced in advance, but to be justified in the end. It could also be treated as a synthesis of these two alternatives, on the model of Marx's unity of research and presentation.

The giving in which Being is given is then determined in accordance with Being as presence:

> Being, by which all beings are marked as such, Being means being present. Considered with regard to what is present, being present shows itself as letting-be-present. But now we must try to consider this letting-be-present explicitly insofar as being present is allowed. Letting-be-present shows its own-most character in bringing into unconcealment. To let be present means to unconceal, to bring into openness. In unconcealing, a giving is at play, that namely which in *letting*-be-present gives the being present, i.e. gives Being.[169]

Metaphysically considered, Being as presence characterizes every being as a presence by letting it be present. Heidegger's question reverses this perspective on Being and asks how this *Being* is given so that it can in turn give beings as present. Metaphysically, that is, Being as presence is a letting-be-*present* (*Anwesen*-lassen) of presences, whereas Heidegger is interested in the way in which Being is *allowed*-to-be-present (Anwesen-*lassen*).

---

[169] *Ibid.*, p. 5, S. 5.

Heidegger has answered this question in various vocabularies and it may be helpful to relate in chart form the terminologies of 1) *SuZ*, 2) *The Thing* and 3) *Time and Being*:

1	beings	Being	Time	Being-itself
2	thing	world	mirroring-play	Appropriation
3	the present being	letting-be-*present*	*letting*-be-present	Appropriation

The vertical dimension indicates the path of negative ontology from categories of the existant to their utopian form of Being and then to a consideration of the relationship of the two forms in terms of the history of Being and its end. The horizontal dimension indicates the two possible directions for an analysis of Being or of letting-be-present: metaphysically to the left or ontologically moving to the right into Appropriation. That a chart cannot faithfully reproduce years of subtle thought or thousands of pages of complex writing need scarcely be stated. Further, the terms which Heidegger uses are inadequate to his task, being part of a metaphysically biased traditional language, and must be understood as "ontic models" which give a hint of something language is incapable of expressing properly.[170] The chart does, however, suggest a relationship between what *SuZ* called Time and what *Time and Being* refers to as the giving of Being, a relationship which warrants exploration.

The terms Time and Being are only the starting point for *Time and Being*, meant to provide a continuity with *SuZ*. In *Time and Being*, Being is re-conceptualized as presence, which is given in a sending as that which is sent in the epochal changes of presence. Presence has a temporal character and so Heidegger investigates Time and the giving of Time to see if it is Time which gives Being, as suggested at *SuZ*'s close. Time is given in the clearing passing of the Time-space in which presence is withheld, given or refused. Thus Time is the giver of Being (or, correspondingly, the withholder or refuser of Being). But what then gives Time? The giver of Being was not found to be Time by some sort of deduction from the phrase "It gives Time," but by uncovering what lay hidden in the characterization of Being as sent presence. Similarly, the giver of Time is to be uncovered in the characterization of Time as clearing passing.[171]

---

[170] Cf. *ibid.*, p. 55f, S. 54f.

[171] *Ibid.*, p. 18f, S. 19f.

The phrases "sent presence" and "clearing passing" indicate a circle of activity. The sending of the presence of temporal beings takes place as the passing of the dimensions of Time. Insofar as one can even distinguish Being and Time any more, they appear together within a mirroring play. In the interplay of Time *and* Being, they are given as clearing passing and sent presence. Time, as clearing passing, thus shows itself to be given in the playful unity of Being and Time, This unity, in which Being and Time appropriate what is appropriate to them, is called by Heidegger *Ereignis*, the event of appropriation.[172]

Just as we concluded earlier in reference to *The Thing* that Appropriation was not something totally separable from the Ring, the Four, the mirroring-play of the world or the thinging of the thing, but that these various abstractions represented a series of successive explications of a single phenomenon, so too in *Time and Being* Appropriation is not a distinct third party to the pair, Being and Time. Rather, Being as *Anwesen*-lassen (presence) reveals itself as Time in the sense of Anwesen-*lassen* (clearing) and the relationship of these two aspects of the same phenomenon is Appropriation, which can be pictured as a non-temporal process of mutual self-appropriation. What smacks of Hegelian mediation cannot strictly speaking be called dialectical, because Being, Time and the event of appropriation are no longer posited as distinct entities which merely presuppose each other. At most they could be considered distinguishable moments of the process in which beings come to be present. In this sense they are not, however, chronologically distinguishable, but conceptual moments in the phenomenological explication of that which "gives" presence (and chronology itself) in its historical modes.

The subtlety of the distinction between Being and Appropriation is captured in Heidegger's statement of purpose: bringing into view Being-itself *as* Appropriation.[173] Throughout Heidegger's writings, the phrase "Being-itself" has been carefully, if not always clearly, distinguished from "Being" and has always referred to what is now called Appropriation.[174] However, the separation of Being as presence from Being-itself as the Appropriation which gives Being is problematized by the use of one word for two functions. Heidegger must have judged it important enough to show the unity of the phenomenon that he used the term "Being" in this plurality of senses, despite the confusion which thereby resulted (e.g. in discussions of the ontological difference).

---

[172] *Ibid.*, p. 19, S. 20.

[173] *Ibid.*, p. 23, S. 24.

[174] Cf. *ibid.*, p. 43, S. 46.

In *The Thing* the unity of *the* ontological phenomenon was expressed in the repetitive phrase, "the world worlds." Although he could not say, "Being is," the term *"Ereignis"* allows Heidegger to repeat this ploy:

> World is present insofar as it worlds. That is, the worlding of world is neither explainable in terms of others nor can it be ground in others.[175]

> What is left to say? Only this: Appropriation   appropriates. With this, we say the same from the same to the same.[176]

But this is the formulation of an ontology of identity, not one seeking an escape from the contemporary form of presence, the danger of our times. Technological objects do not "thing" because this world does not "world," because Appropriation does not "appropriate." In the sending of Being and the passing of Time, Appropriation retreats and withholds or refuses itself: Appropriation expropriates itself (*enteignet*).[177]

Negative ontology is one of finitude. In *SuZ*, Dasein's finitude is expressed in the limitation of his possibilities to be, imposed by his birth, situation and death, and by his limited understanding of these possibilities as his own. The finitude of being is manifold: presence is always given in one form and thereby not in others (although at any moment there may be an overlapping of epochs as seen in the contemporaneity of van Gogh's painting with the shoes it pictures, or individual phenomena may be subject to a plurality of interpretations, e.g. by a philosopher, an art critic and a museum custodian). Time involves the absence of presence withheld or refused. Appropriation keeps to itself in the sending of Being and the passing of Time. Appropriation reveals itself only negatively, as Expropriation, in its absence from the sending of Being and the passing of Time, as the forgotten question of Being, as the danger of our times and as the task for critical thought.

The Concept of Being

In the preceding, we have tried to sketch-in a coherent, more or less intelligible system of thought as an interpretation of Heidegger's central writings, of the path and thrust of his thinking about Being. We have tried to

---

[175] "The Thing," S. 52.

[176] "Time and Being," p. 24, S. 24.

[177] *Ibid.*, p. 22, S. 23.

understand his negative ontology – an attempt to grasp the way in which beings are given as present without thereby setting contemporary Being as an absolute – from the perspective of our historical ontological question: how it comes to be that a certain sense of Being prevails in a given era. Assuming that what Heidegger has said is valid and that we have understood him correctly, and granting that he has presented a wealth of material which we have ignored, we can now pose the question: Has Heidegger in the end offered us anything substantial under his cloud of analyses?

Heidegger's path of thought has led him to criticize the entire philosophical tradition from the pre-Socratics to Husserl, as a whole and in particulars. He has also proposed powerful, original perspectives for answering the eternal philosophical questions concerning space, time and Being; art, science and the humanities; knowledge, truth and thought; things, tools, works and mortals; history, freedom and death. Yet, in the end the question forces itself upon us, as upon Heidegger himself: "But do we arrive by this road at anything other than to a mere mental construct?"[178] Is Heidegger's thought so deep that it bypasses all content, so abstract that it has no concrete significance, so essential that it forgets its original inspiration: that essence lies in the particular existence? We have seen that originally (in *SuZ*) the abstractness of the answer to the question of Being was to be counteracted by the concrete destruction of the history of ontology. In *Time and Being*, these concrete analyses have already been incorporated into the discussion – to little avail. The seminar to *Time and Being* touched upon the problems of non-specificity several times.

In the seminar, Heidegger was asked if it was sufficient to grasp the relation of presence to what is present as uncovering (*Entbergen*) when this term is taken in abstraction from all content: "If unconcealing already lies in all kinds of *poiesis*, of making, of effecting, how can one exclude these modes and keep unconcealing purely for itself?"[179] To this question, Heidegger responded that unconcealing (*Entbergen*) in the sense he used it was in fact more general than e.g. Plato's usage of *poiesis* because he referred to the uncovering of the whole being as such, not e.g. just to its *eidos*, *what* it is as distinct from *that* it is. He had to admit, however, that it remained a task of thought to determine the uncoveredness of the different realms of things.[180]

The concept of presence presents the same problem. Heidegger finds it in all the metaphysical conceptions of Being; if his lecture ever seemed to proceed

---

[178] *Ibid.*, p. 23, S. 24.

[179] *Ibid.*, p. 46f, S. 50.

[180] *Ibid.*, p. 47, S. 50f.

deductively, it was from the characterization of Being in all its forms as presence, with its temporal connotation. Yet, Heidegger must admit, the first principle of Being as presence is totally questionable: "The priority of presence thus remains an assertion in the lecture *Time and Being*, but as such a *question* and a task of thinking, namely to consider whether and whence and to what extent the priority of presence exists."[181] These terms, uncovering and presence, are no exceptions among Heidegger's tangle of concepts. Leaving the dirty work for one's followers is not simply a lazy man's trick; it is questionable whether the generality of a concept is an asset, especially if it is not supported by a pyramid of more specific concepts applicable to the various aspects as well as the different regions. Heidegger's specific analyses of *Angst*, the work of art or the jug may have been brilliant in their day or they may have been trite and absurd – they were not meant to be judged on their own. Whatever their value, they could never justify by themselves the generalizing leaps that followed in their wakes; such generalizations may in the end be unjustifiable or lead to mere emptiness.

The emptiness of Heidegger's generality is perhaps most painful in connection with his attempt at social criticism. In contrast to the insightful and subtle handling of complex and abstruse considerations throughout the seminar, the blatantly naive question about how the technological sense of Being is limited to our planet was ignored. "It was not explained how the *Gestell*, which constitutes the essence of modern technology, hence of something that, as far as we know, only occurs on earth, can be a name for universal Being."[182] A confusion about the scope of the problem might not be so disastrous if Heidegger were clearer about the nature of the salvation from the technological danger. Alas, here he is not even sure what he does not know. He remains unshaken in his confidence *that* he can reach salvation by stepping ever backwards into increasing abstraction, but where, what or how salvation will be reached he cannot imagine.

> To the extent to which this was clarified, one could say in spite of the inadequacy of these expressions: The 'that' of the place of the 'whither' is certain, but as yet how this place is, is concealed from knowledge. And it must remain undecided whether the 'how', the manner of Being of this place, is already determined (but not yet knowable) or whether it itself results only from the taking of the step, in the awakening into Appropriation.[183]

---

[181] *Ibid.*, p. 34, S. 37.

[182] *Ibid.*, p. 33, S. 35.

[183] *Ibid.*, p. 30f, S. 33.

Not that Heidegger has to have all the answers, but his poverty at the crucial point raises further questions. How does he know that what he is searching for is to be found if he knows nothing about it? Is this merely wishful thinking done up in fancy jargon? If the general concepts remain questions and the concrete ground-work has yet to be carried out, does Heidegger really have more to offer, even in the form of an ontological foundation, than Marx?

Concluding Remarks

Both Marx and Heidegger formulate theories of technological Being, expressed in the related conceptions of abstract value and calculable stock. For each of the thinkers, the theory of technological society is elaborated within an historical meta-ontology, which attempts to comprehend the contemporary form of Being as having developed out of Western civilization and to criticize it as limited, contradictory and self-concealing. But, whereas man, beings and Being-itself are treated by Heidegger as if they were monads with windows to each other but no developed relations, Marx grasps them precisely by their mediations. Heidegger, claiming to inquire after the conditions of the possibility of their having relations to each other, hypostatizes even Being – which is no being, but a moment in the mediation of beings – into an in-itself with essential characteristics, possibilities and temporality. Marx, in contrast, understands people and their products as determining the totality of interrelations, which in turn determines them, a totality which is most appropriately conceptualized by a theory of the mode of production as the primary sphere of mediation. The term "Being" is unnecessary to Marx's theory for it is implicitly dealt with, rather than being fixated upon and glorified.

For Heidegger, as for Hegel before him, the developmental process whereby Being, which determines the form of presence of beings, is itself determined takes place solely within the realm of Being-itself. In Marx's theory, on the contrary, the history of Being is the consequence of concrete human history, and its apparent autonomy from human control is an illusion resulting from the complexity of historical mediations within an antagonistically structured society. Marx's ontological essences, above all that of abstract value, are accordingly derived from concrete, historically-specific categories, such as exchange value, comprehended as the form of appearance of the essence. Actual beings are thus not simply objectifications or placeholders of a Being that develops independently; the history of Being is not a mystical intergalactic happening or even a process taking place primarily within the language of a people or the intellectual history of a tradition. *That beings are now present as calculable stock, abstracted from their unique context and physical characteristics, is, according to Marx, primarily a result of their being present in relations of exchange. It is these concrete relations of beings to beings as they have developed in social, economic, material history, which equate the forces used in the production of each commodity with all other forces of production, equate each being with every other commodity, equate the human activity involved in any task with labor as such,* and thereby abstract from the mortality and situatedness of people.

Marx thus understands the prevailing form of presence in relation to the social totality, whose character is essentially conditioned by the prevalent mode of production. For Marx, history progresses through a dialectic of whole and part, of social production and its various products. Heidegger, however, investigating the preconditions of this process, loses sight of the dialectical relationship in favor of a one-sided determination by Being of the form of presence of beings. Where Marx understands the preconditions of one epoch as the conditions of its predecessor, Heidegger accepts the character of an epoch as fatefully given and beyond comprehension. *The triviality of Heidegger's social commentary in comparison to Marxian social analysis is thus neither accidental nor is it to be enriched through the addition of concrete details. Being, which determines beings as beings, must itself be shown to be conditioned by beings.* The ontological self-interpretation of the world is not divorced from the ontic self-transformation of the world; thought which attempts to comprehend the former cannot ignore its unity with the latter, as Heidegger does.

Bibliography

Adorno, Theodor W. *The Jargon of Authenticity*. Evanston: Northwestern University Press. 1973.

Adorno, Theodor W. Letter to Walter Benjamin dated 10 November 1938. *New Left Review*. No. 81. October 1973.

Adorno, Theodor W. *Negative Dialectics*. New York: Seabury. 1973.

Adorno, Theodor W. A portrait of Walter Benjamin. *Prisms*. London: Neville Spearman. 1967.

Adorno, Theodor W. Thesen über die Sprache des Philosophen. *Gesammelte Schriften I*. Frankfurt: Suhrkamp. 1973.

Adorno, Theodor W. Der wunderliche Realist: Über Siegfried Kracauer. *Noten zur Literatur VI*. Frankfurt: Suhrkamp. 1965.

Apel, Karl-Otto. *Transformation der Philosophie*. 2 vols. Frankfurt: Suhrkamp. 1973.

Benjamin, Walter. *Illuminations*. New York: Schocken. 1969.

Continuum. Vol. 8. Nos. 1 & 2. Chicago. 1970.

Enzensberger, Hans-Magnum. Critique of political ecology. *New Left Review*. No. 84.

Gadamer, Hans-Georg. *Wahrheit und Methode: Grundzüge einer philosophischen Hermeneutik*. 2nd. ed. Tübingen: Mohr. 1965.

Habermas, Jürgen. *Knowledge and Human Interests*. Boston: Beacon Press. 1971.

Habermas, Jürgen. Zur Logik der Sozialwissenschaften. *Philosophische Rundschau*. Beiheft 5. Tübingen. February 1967.

Heidegger, Martin. Bauen Wohnen Denken. *Vorträge und Aufsätze*. Bd. II. Tübingen: Neske, 1954.

Heidegger, Martin. Dichterisch Wohnet der Mensch. *Vorträge und Aufsätze*. Bd. II. Tübingen: Neske. 1954.

Heidegger, Martin. Das Ding. *Vorträge und Aufsätze*. Bd. II. Tübingen: Neske. 1954.

Heidegger, Martin. *Einführung in die Metaphysik*. Tübingen: Neske. 1958.

Heidegger, Martin. *An Introduction to Metaphysics*. Garden City: Anchor. 1961.

Heidegger, Martin. Das Ende der Philosophie und die Aufgabe des Denkens. *Zur Sache des Denkens*. Tübingen: Niemeyer. 1969.

Heidegger, Martin. The end of philosophy and the task of thinking. *On Time and Being*. New York: Harper & Row. 1972.

Heidegger, Martin. Der Feldweg. *Martin Heidegger zum 80. Geburtstag*. Frankfurt: Klostermann. 1969.

Heidegger, Martin. Hegel und die Griechen. *Wegmarken.* Frankfurt: Klostermann. 1967.

Heidegger, Martin. Humanismusbrief. *Wegmarken.* Frankfurt: Klostermann. 1967.

Heidegger, Martin. Letter on humanism. *Philosophy in the Twentieth Century.* Vol. III. New York: Harper & Row. 1971.

Heidegger, Martin. Kants These über das Sein. *Wegmarken.* Frankfurt: Klostermann. 1967.

Heidegger, Martin. *Nietzsche.* Bd. II. Pfüllingen: Neske. 1961.

Heidegger, Martin. *The Question of Being.* Bilingual ed. New Haven: College & University Press. 1958.

Heidegger, Martin. *Raum und Kunst.* St. Galen: Erker Verlag. 1969.

Heidegger, Martin. *Der Satz von Grund.* Pfüllingen: Neske. 1971.

Heidegger, Martin. *Sein unt Zeit.* Tübingen: Niemeyer. 1967.

Heidegger, Martin. *Being and Time.* New York: Harper& Row. 1962.

Heidegger, Martin. Der Ursprung des Kunstwerkes. *Holzwege.* Frankfurt: Klostermann. 1963.

Heidegger, Martin. The origin of the work of art. *Philosophies of Art and Beauty.* New York: Modern Library. 1964.

Heidegger, Martin. *Der Ursprung des Kunstwerkes.* Stuttgart: Reclam. 1960.

Heidegger, Martin. Vom Geheimnis des Glockenturms. *Martin Heidegger zum 80. Geburtstag.* Frankfurt: Klostermann. 1969.

Heidegger, Martin. *Was Heisst Denken?* Tübingen: Niemeyer. 1971.

Heidegger, Martin. *What is Called Thinking?* New York: Harper & Row. 1968.

Heidegger, Martin. Was ist Metaphysik? *Wegmarken.* Frankfurt: Klostermann. 1967.

Heidegger, Martin. What is metaphysics? *Existence and Being.* Chicago: Gateway. 1965.

Heidegger, Martin. *What is Philosophy?* Bilingual ed. New Haven: College & University Press. n.t.

Heidegger, Martin. Zeit und Sein. *Zur Sache des Denkens.* Tübingen: Niemeyer. 1969.

Heidegger, Martin. Time and Being. *On Time and Being.* New York: Harper & Row. 1972.

Hermeneutik und Ideologiekritik. Frankfurt: Suhrkamp. 1971.

Horkheimer, Max & Adorno, Theodor W. *Dialectic of Enlightenment.* New York: Herder & Herder. 1972.

Husserl, Edmund. Der Ursprung der Geometrie. *Die Krisis der europäischen Wissenschaften und die transzendentale Phänomenologie.* The Hague: Martinus Nijhoff. 1954.

Jünger, Ernst. *Der Arbeiter: Herrschaft und Gestalt.* Hamburg. 1932.

Lukacs, Georg. *Geschichte und Klassenbewusstsein.* Berlin: Malik. 1967.

Marcuse, Herbert. *One-Dimensional Man.* Boston: Beacon. 1964.

Marcuse, Herbert. The affirmative character of culture. *Negations*. Boston: Beacon. 1969.

Martin Heidegger im Gespräch. Freiburg: Alber. 1969.

Marx, Karl. *A Contribution to the Critique of Political Economy*. New York: International. 1970.

Marx, Karl. Einleitung. *Grundrisse der Kritik der politischen Ökonomie (Rohentwurf) 1857-1858*. Photocopy of 1939. Moscow editions. Frankfurt: Europische Verlagsanstalt. 1941.

Marx, Karl. Introduction. *Grundrisse*. London: Penguin. 1973.

Marx, Karl. *Pre-Capitalist Economic Formations*. New York: International Publishers. 1969.

Marx, Karl. *Das Kapital*, Bd. I. Frankfurt: Ullstein. 1971.

Marx, Karl. *Capital*. Vol. I. New York: International. 1967.

Marx, Karl. Ökonomisch-Philosophische Manuskripten (1844). *Texte zu Methode und Praxis*. Bd. II. Reinbek: Rowohlt. 1966.

Marx, Karl. Economic and philosophic manuscripts (1844). *Writings of the Young Marx on Philosophy and Society*. Garden City: Doubleday, 1967.

Petrovik, Gajo. Der Spruch des Heideggers. *Durchblicke. Martin Heidegger zum 80. Geburtstag*. Frankfurt: Klostermann. 1970. S. 412ff.

Pöggeler, Otto. *Philosophie und Politik bei Heidegger*. Freiburg: Alber. 1972.

Schwann, Alexander. *Politische Philosophie im Denken Heideggers*. Koln: Westdeutscher. 1965.

Schmidt, Alfred. *Marx's Concept of Nature*. London: New Left Books. 1972.

Schmidt, Alfred. *Geschichte und Struktur*. Munchen: Hauser. 1971.

Vita   .

1945   Born at the dawn of advanced industrial capitalism on March 16 near Wilmington, Delaware, into a family committed to trade unionism and socialist liberalism.

1963-67   At M.I.T., discovered limitations of the modern scientistic perspective and sought acceptable critical alternatives. Introduced to phenomenology and existentialism. Awarded S.B. in philosophy and mathematics. Received political education in the New Left.

1967-68   Explored Europe and continental thought while honeymooning in romantic Heidelberg. Concentrated on Gadamer's hermeneutics and Heidegger's ontology.

1968-70   Taught high school, drove a cab, did computer systems programming, helped raise a son. Took several philosophy courses at Temple U.

1970-71   During graduate study at Northwestern U., organized study groups on Marx, hermeneutics and Habermas, translating some of the texts. Was granted M.A. and permitted to undertake research into the meeting-grounds of Marxism and existentialism.

1971-73	Attended University of Frankfurt/Main, studying Hegel, Marx, Adorno, Habermas. Followed discussion within remnants of the *Institut für Socialforschung*. Wrote dissertation chapters on Heidegger.
1973-74	Taught courses on Marx, Wittgenstein, Heidegger, Adorno at Evening Division of Northwestern U. Organized study groups on Frankfurt School. Developed key aspects of the Marx interpretation; composed and clarified bulk of the dissertation.
1974-75	Returned to computer work and joined the organizing drive for an AFSCME union local. Experimented in electronic music composition, suburban living and political discourse. Analyzed opening dialectic of *Das Kapital* and formulated dissertation's prefatory, transitional and concluding remarks. Declared dissertation complete on thirtieth birthday and defended it May 8. Published chapter on Adorno's Heidegger critique in *Boundary II*, a journal of postmodern literature.

www.ingramcontent.com/pod-product-compliance
Lightning Source LLC
Chambersburg PA
CBHW032010170526
45157CB00002B/625